B. PANKOW

PRESSURE
VESSEL
DESIGN
HANDBOOK

PRESSURE VESSEL DESIGN HANDBOOK

Henry H. Bednar, P.E.

VNR **VAN NOSTRAND REINHOLD COMPANY**
NEW YORK   CINCINNATI   ATLANTA   DALLAS   SAN FRANCISCO
LONDON   TORONTO   MELBOURNE

Van Nostrand Reinhold Company Regional Offices:
New York   Cincinnati   Atlanta   Dallas   San Francisco

Van Nostrand Reinhold Company International Offices:
London   Toronto   Melbourne

Library of Congress Catalog Card Number: 81-1278
ISBN: 0-442-25416-4

Manufactured in the United States of America

Published by Van Nostrand Reinhold Company
135 West 50th Street, New York, N.Y. 10020

Published simultaneously in Canada by Van Nostrand Reinhold Ltd.

15 14 13 12 11 10 9 8 7 6 5 4 3 2 1

Library of Congress Cataloging in Publication Data
Bednar, Henry H.
   Pressure vessel design handbook.

   Includes bibliographical references and index.
   1. Pressure vessels—Design and construction—
Handbooks, manuals, etc. I. Title.
TS283.B42   681'.76041   81-1278
ISBN 0-442-25416-4   AACR2

Preface

This handbook has been prepared as a practical aid for engineers who are engaged in the design of pressure vessels. Design of pressure vessels has to be done in accord with specific codes which give the formulas and rules for satisfactory and safe construction of the main vessel components. However, the codes leave it up to the designer to choose what methods he will use to solve many design problems; in this way, he is not prevented from using the latest accepted engineering analytical procedures.

Efficiency in design work is based on many factors, including scientific training, sound engineering judgement, familiarity with empirical data, knowledge of design codes and standards, experience gained over the years, and available technical information. Much of the technical information currently used in the design of pressure vessels is scattered among many publications and is not available in the standard textbooks on the strength of materials.

This book covers most of the procedures required in practical vessel design. Solutions to the design problems are based on references given here, and have been proven by long-time use; examples are presented as they are encountered in practice. Unfortunately, exact analytical solutions for a number of problems are not known at the present time and practical compromises have to be made.

Most engineering offices have developed their own vessel calculation procedures, most of them computerized. However, it is hoped that this book will provide the designer with alternative economical design techniques, will contribute to his better understanding of the design methods in use, and will be convenient when hand computations or verifications of computer-generated results have to be made.

No particular system of notation has been adopted. Usually the symbols as they appear in particular technical sources are used and defined as they occur. Only the most important references are given for more detailed study.

It is assumed that the reader has a working knowledge of the ASME Boiler and Pressure Vessel Code, Section VIII, Pressure Vessels, Division 1.

The writer wishes to express his appreciation to the societies and companies for permission to use their published material.

Finally, the writer also wishes to express his thanks to the editorial and production staff of the Publisher for their contribution to a successful completion of this book.

HENRY H. BEDNAR

Contents

PRESSURE
VESSEL
DESIGN
HANDBOOK

1
Design Loads

1.1. INTRODUCTION

The forces applied to a vessel or its structural attachments are referred to as loads and, as in any mechanical design, the first requirement in vessel design is to determine the actual values of the loads and the conditions to which the vessel will be subjected in operation. These are determined on the basis of past experience, design codes, calculations, or testing.

A design engineer should determine conditions and all pertaining data as thoroughly and accurately as possible, and be rather conservative. The principal loads to be considered in the design of pressure vessels are:

design pressure (internal or external),
dead loads,
wind loads,
earthquake loads,
temperature loads,
piping loads,
impact or cyclic loads.

Many different combinations of the above loadings are possible; the designer must select the most probable combination of simultaneous loads for an economical and safe design.

Generally, failures of pressure vessels can be traced to one of the following areas:

material: improper selection for the service environment; defects, such as inclusions or laminations; inadequate quality control;

design: incorrect design conditions; carelessly prepared engineering computations and specifications; oversimplified design computations in the absence of available correct analytical solutions; and inadequate shop testing;

fabrication: improper or insufficient fabrication procedures; inadequate inspection; careless handling of special materials such as stainless steels;

1

service: change of service conditions to more severe ones without adequate provision; inexperienced maintenance personnel; inadequate inspection for corrosion.

1.2. DESIGN PRESSURE

Design pressure is the pressure used to determine the minimum required thickness of each vessel shell component, and denotes the difference between the internal and the external pressures (usually the design and the atmospheric pressures—see Fig. 1.1). It includes a suitable margin above the operating pressure (10 percent of operating pressure or 10 psi minimum) plus any static head of the operating liquid. Minimum design pressure for a Code nonvacuum vessel is 15 psi. For smaller design pressures the Code stamping is not required. Vessels with negative gauge operating pressure are generally designed for full vacuum.

The *maximum allowable working (operating) pressure* is then, by the Code definition, the maximum gauge pressure permissible at the top of the completed vessel in its operating position at the designated temperature. It is based on the nominal vessel thickness, exclusive of corrosion allowance, and the thickness required for other loads than pressure. In most cases it will be equal or very close to the design pressure of the vessel components.

By the Code definition, the *required thickness* is the minimum vessel wall thickness as computed by the Code formulas, not including corrosion allowance; the *design thickness* is the minimum required thickness plus the corrosion allowance; the *nominal thickness* is the rounded-up design thickness as actually used in building the vessel from commercially available material.

If the nominal vessel thickness minus corrosion allowance is larger than the required thickness, either the design pressure or the corrosion allowance can be

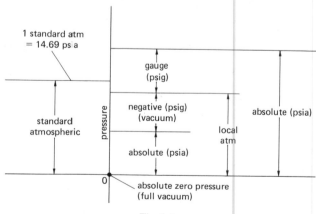

Fig. 1.1.

increased, or any excess thickness can be used as reinforcement of the nozzle openings in the vessel wall.

The vessel shell must be designed to withstand the most severe combination of coincident pressure and temperature under expected operating conditions. The *nominal stress* in any part of the vessel as computed from the Code and standard engineering stress formulas, without consideration of large increases in stresses at the gross structural discontinuities, cannot exceed the Code allowable stress.

1.3. DESIGN TEMPERATURE

Design temperature is more a design environmental condition than a design load, since only a temperature change combined with some body restraint or certain temperature gradients will originate thermal stresses. However, it is an important design condition that influences to a great degree the suitability of the selected material for construction. Decrease in metal strength with rising temperature, increased brittleness with falling temperature, and the accompanying dimensional changes are just a few of the phenomena to be taken into account for the design.

The required Code design temperature is not less than the mean metal vessel wall temperature expected under operating conditions and computed by standard heat transfer formulas and, if possible, supplemented by actual measurements. For *most standard vessels* the design temperature is the maximum temperature of the operating fluid plus 50°F as a safety margin, or the minimum temperature of the operating fluid, if the vessel is designed for low-temperature service (below -20°F).

In large process vessels such as oil refinery vacuum towers the temperature of the operating fluid varies to a large degree, and zones with different design temperatures, based on expected calculated operating conditions, can be used for the design computations of the required thicknesses.

The design metal temperature for *internally insulated vessels* is determined by heat transfer computations, which should provide sufficient allowance to take care of the probable future increase in conductivity of the refractory due to gas, deterioration, coking, etc. At a minimum, the designer should assume a conductivity for the internal insulating material 50–100 percent higher than that given by the manufacturer's data, depending on operating conditions. A greater temperature margin should be used when external insulation is used as well. The possibility of a loss of a sizable lining section and the required rupture time of the shell should also be considered. Extensive temperature instrumentation of the vessel wall is usually provided.

For *shut-down conditions* the maximum design temperature for uninsulated vessels and the connecting piping will be the equilibrium temperature for metal objects, approximately 230°F for the torrid zone, 190°F for the temperate zone, and 150°F for the frigid zone.

The lowest design metal temperature *for pressure storage vessels* should be taken as 15°F above the lowest one-day mean ambient temperature for the particular location.

The design temperature for *flange through bolts* is usually lower than the temperature of the operating fluid, unless insulated, and it can be safely assumed to be 80 percent of the design vessel temperature. However, the external tap and the internal bolting should have the same design temperature as the vessel.

When the design computations are based on the thickness of the base plate exclusive of lining or cladding thickness, the maximum service metal temperature of the vessel should be that allowed for the base plate material.

The design temperature of *vessel internals* is the maximum temperature of the operating liquid.

1.4. DEAD LOADS

Dead loads are the loads due to the weight of the vessel itself and any part permanently connected with the vessel. Depending on the overall state, a vessel can have three different weights important enough to be considered in the design.

1. *Erection (empty) dead load* of the vessel is the weight of the vessel without any external insulation, fireproofing, operating contents, or any external structural attachments and piping. Basically, it is the weight of a stripped vessel as hoisted on the job site. In some small-diameter columns the removable internals (trays) are shop-installed, and they have to be included in the erection weight. Each such case has to be investigated separately.

2. *Operating dead load* of the vessel is the weight of the in-place completed vessel in full operation. It is the weight of the vessel plus internal or external insulation, fireproofing, all internals (trays, demister, packing, etc.) with operating liquid, sections of process piping supported by the vessel, all structural equipment required for the vessel servicing and inspection (platforms, ladders, permanent trolleys etc.), and any other process equipment (heat exchangers) attached to the vessel.

3. *Shop test dead load* of the vessel consists only of the weight of the vessel shell, after all welding is finished, filled with test liquid.

Field test dead load is the operating dead load with only external or/and internal insulation removed for inspection purposes and filled fully with the test liquid instead of operating liquid. This load is used as a design load only if the vessel is expected to be tested in field at some future date.

The ice or snow load as well as any live load (weight of the maintenance personnel with portable tools) is considered to be negligible.

1.5. WIND LOADS

Wind can be described as a highly turbulent flow of air sweeping over the earth's surface with a variable velocity, in gusts rather than in a steady flow. The wind can also be assumed to possess a certain mean velocity on which local three-dimensional turbulent fluctuations are superimposed. The direction of the flow is usually horizontal; however, it may possess a vertical component when passing over a surface obstacle. The wind velocity V is affected by the earth surface friction and increases with the height above the ground to some maximum velocity at a certain gradient level above which the wind velocity remains constant. The shape of the velocity profile above the ground depends on the roughness characteristics of the terrain, such as flat open country, wooded hilly country-side, or a large city center. It can be expressed by the power-law formula

$$V \propto z^n$$

where the value of the exponent n depends on the terrain and z is the elevation above the ground level.

Since the standard height for wind-speed recording instruments is 30 ft, the power formula is used to correct standard-height velocity readings for other heights above ground with any given terrain profile (see Fig. 1.2).

The velocity (dynamic) pressure representing the total kinetic energy of the moving air mass unit at the height of 30 ft above the ground on a flat surface perpendicular to the wind velocity is given by the equation:

$$q_{30} = \rho V^2/2 = (\tfrac{1}{2})0.00238(5280/3600)^2 V_{30}^2 = 0.00256 V_{30}^2$$

where

V_{30} = basic wind speed at 30 ft height above the ground in mph, maximum as measured on the location over a certain period of time

q_{30} = basic wind velocity pressure at 30 ft height above the ground in psf

ρ = 0.00238 slugs per ft^3, air density at atm pressure, and 60°F.

The magnitude of the basic wind velocity V_{30} used in determination of the design pressure q_{30} depends on the geographical location of the job site. The wind pressure q_{30} is used to compute the actual wind design loads on pressure vessels and connected equipment. However, since the wind velocity V is influenced by the height above the ground and terrain roughness and the pressure q itself is influenced by the shape of the structure, the basic wind velocity pressure q_{30} has to be modified for different heights above the ground level and different shapes of structures.

In doing so either the older, somewhat simpler standard ASA A58.1-1955 or the new revised ANSI A58.1-1972 are generally used, unless the client's specifi-

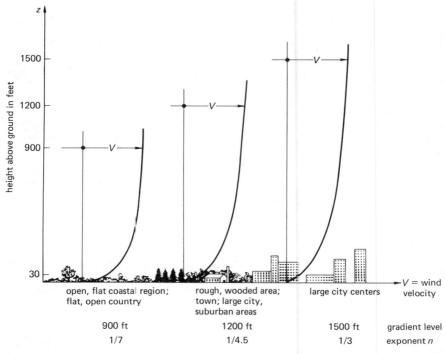

Fig. 1.2. Wind velocity profiles over three basic terrain roughness characteristics. If the wind velocity in flat, open country at the standard height 30 ft is $V_{30} = 60$ mph, at the gradient level 900 ft the wind velocity is

$$V = 60(900/30)^{1/7} \doteq 100 \text{ mph.}$$

In the same region, the gradient wind velocity in the suburban area is the same, i.e., $V = 100$ mph, and the gradient level is 1200 ft; hence the wind velocity at the standard level in the suburban area is

$$V_{30} = 100(30/1200)^{1/4.5} = 44 \text{ mph.}$$

cations dictate otherwise. Although the former standard is obsolete, it was extensively used for many years and it is still used in some codes. It is therefore quite probable that for some time to come the designer will encounter designs where the wind loads were computed on the basis of the old standard. Also the designer may become engaged in the construction of pressure vessels for petrochemical plants in foreign countries where long-term wind velocity data are lacking and the design procedures are specified more in line with the now obsolete specification.

Wind Loads as Computed in Accordance with ASA Specification A58.1-1955

The procedure for calculation of the minimum design wind load normal to the surface of the structure is as follows.

1. The geographical area of the job site is located on the wind pressure map, (see Fig. 1 of the standard). The basic wind pressure p is selected. It is based on the maximum regional measured wind velocity V_{30} (excluding tornado velocities) and includes the shape factor 1.3 for flat surface and the gust factor 1.3 for heights up to 500 ft above the ground. No distinction is made for terrain roughness or mean recurrence interval of the highest wind in the area.

2. The wind design pressures p_z, corresponding to the basic wind pressure p, for various height zones above the ground are given in Table 3 of the standard. They include a height factor based on the seventh-root law to express the variation of the wind velocity with the height above the ground.

3. The shape factor B for round objects is equal to 0.6 and is applied to the design pressure p_z. The shape factors, as they appear in the standard are actual shape coefficients divided by the flat surface factor 1.3 which was included in the basic wind pressure p.

4. If the windward surface area projected on the vertical plane normal to the direction of the wind is A ft^2, then the resultant of the wind pressure load over the area P_w is assumed to act at the area centroid and is given by

$$P_w = ABp_z \text{ lb.}$$

The wind pressure forces are applied simultaneously, normal to all exposed windward surfaces of the structure. The minimum net pressure (Bp_z) in the above formula for cylindrical vertical vessels is not less than 13 psf for $L/D \leqslant 10$ and 18 psf for $L/D \geqslant 10$, where L is the overall tangent-to-tangent length of the vessel and D is the vessel nominal diameter.

Wind Loads as Computed in Accordance with ANSI A58.1-1972

The effective wind velocity pressures on structures q_F and on parts of structures q_p in psf at different heights above the ground are computed by the following equations:

$$q_F = K_z G_F q_{30}$$
$$q_p = K_z G_p q_{30}$$

where

K_z = velocity pressure coefficient depending on the type of the terrain surface (exposure) and the height above the ground z. To simplify computation, height zones of constant wind velocity can be assumed. For instance the pressure q_F at 100 ft above the ground can be applied over the zone extending from 75 to 125 ft above the ground.

G_F = gust response factor for structure

G_p = gust response factor for portion of structures.

The variable gust response factors were introduced to include the effect of sudden short-term rises in wind velocities. The values of G_F and G_p depend on the type of the terrain exposure and dynamic characteristics of structures. For tall cylindrical columns, for which gust action may become significant, detailed computations of G_F are usually required as shown in the Appendix to the standard. Since the equation for G_F contains values which depend on the first natural frequency f of the column, which in turn depends on the vessel wall thickness t to be computed from the combined loadings (wind, weight, and pressure), the additional mathematical work involved in successive approximation may render this standard less attractive than the previous one.

The q_F and q_p values have to be further modified by a net pressure coefficient C_f for different geometric shapes of structures.

If the projected windward area of the vessel section on a vertical plane normal to the wind direction is A ft^2, the total design wind load P_w on a vessel section may be computed by equation:

$$P_w = AC_f q_F \text{ lb}$$

The minimum net pressure $(C_f q_F)$ in the above formula should be not less than 15 psf for the design of structures and 13 psf for structural frames. The wind loads on large appurtenances such as top platforms with trolley beams, piping, etc. can be computed in the same manner, using appropriate C_f and q_F with allowances for shielding effect and must be added to the wind load acting on the entire vessel.

The three standard terrain roughness categories selected are as shown in Fig. 1.2.

a. Exposure A: centers of large cities, rough and hilly terrains;
b. Exposure B: rolling terrains, wooded areas, suburban areas;
c. Exposure C: flat, open grass country, coastal areas.

Most large petrochemical plants will belong to category C.

The procedure for computing the minimum design wind load on an enclosed structure such as a tall column can be summarized as follows.

1. Using as criteria the anticipated service life of the vessel and the magnitude of the possible damage in case of failure, the basic wind speed V_{30} is selected *from Fig. 1 or 2 of the standard* for the particular job location and modified by special local conditions; see also Appendix A1 of this book.

2. The basic wind pressure $q_{30} = 0.00256 V^2$ is computed.

3. The effective wind velocity pressure q_F is given by $q_F = K_z G_F q_{30}$. The height zones of constant wind velocities are selected and K_z determined from Fig. A2 of the standard for each zone. The gust response factor G_F, which does

not change with the height above the ground, is computed from the equation $G_F = 0.65 + 1.95\,(\sigma\sqrt{P})$.

4. Using the net pressure coefficient C_f from Table 15 of the standard and the projected area A in ft^2, the total wind load on the vessel sections of constant wind design pressure may be evaluated by equation $P_w = C_f A q_F$.

Computation of the Projected Area A

It is not possible for the designer to evaluate the projected windward area A of a tower and all appurtenances accurately. When a vessel is being designed only the main features such as the inside diameter, overall length, nozzle sizes, number of manholes, etc. are known and a complete layout is unavailable.

To arrive to some reasonable approximation of the projected area, some assumptions based on past experience must be made. An approximate layout sketch of the vessel with all probable platforms, ladders, and connected piping can be made and with resulting wind loads, wind shears and wind moments at different heights above the ground can be computed. Unless the vessel is comparatively simple, for instance a short vertical drum with a top platform as in Fig. 5.8, this approach is time-consuming and not really justified.

An approach to computing A which is often used and is recommended here is to increase the vessel diameter D to the so-called effective vessel diameter to approximate the combined design wind load:

$$D_e = (\text{vessel o.d.} + \text{twice insulation thickness}) \times K_d.$$

The coefficient K_d is given in Table 1.1. The required projected area A will then be equal to

$$A = D_e \times H_s$$

where H_s = length of the shell section in the zone of the uniform wind velocity. However, the effective diameter D_e can be derived by a simple procedure which allows the designer to adjust D_e according to the actual standard vessel layout.

Table 1.1.

VESSEL OUTSIDE DIAMETER INCLUDING INSULATION	COEFFICIENT K_d
less than 36 in.	1.50
36 to 60 in.	1.40
60 to 84 in.	1.30
84 to 108 in.	1.20
over 108 in.	1.18

Source: Ref. 6.

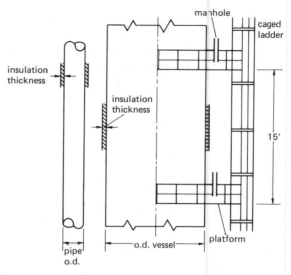

Fig. 1.3. Assumed column layout for determination of the effective diameter D_e.

According to an assumed typical section of a process column shown in Fig. 1.3 the principle parts contributing to the total wind load are as follows:

1. Vessel shell outside diameter with twice the insulation thickness, if any.
2. Adjusted platform area. Assuming half of the platform, 3 ft 6 in. wide at each manhole, spaced at 15 ft, the equivalent increase in the vessel diameter $(42 \times 18)/(15 \times 12) = 4.2$ in.
3. Caged ladder. Assume one caged ladder running from the top of the vessel to the ground. The increase in the column diameter can be taken as 12 in.
4. Piping. Assume the largest pipe in the top third of the column running to the ground level.

All the above items can and should be adjusted according to the actual standard layout as used.

For example, the effective diameter of a 6-ft-diameter column with 1-in. wall thickness, 2-in. insulation, and a 6-in. nozzle in the top third with 1-in. insulation is computed as follows:

$$D_e = (\text{vessel o.d.} + 2 \times \text{insulation thickness})$$
$$+ (\text{pipe o.d.} + 2 \times \text{insulation thickness}) + (\text{platform}) + (\text{ladder})$$
$$= (74 + 4) + (6.625 + 2) + 4.2 + 12 = 102.8 \text{ in.} = 8.6 \text{ ft}$$

The factor $K_d = 102.8/78 = 1.31$ is in agreement with the value in Table 1.1.

The formula above does not include special attached equipment such as heat exchangers or large-top oversized platform with lifting equipment, whose wind loads and moments are computed separately and added to the above.

Example 1.1. Determine the wind loads acting on the process column shown in Fig. 1.4 with an average wall thickness of 1 in., insulation thickness 1.5 in., located in the vicinity of Houston, Texas, using:

   a. ASA A58.1-1955
   b. ANSI A58.1-1972.

   a. Effective diameter $D_e = K_d \times$ o.d. $= 1.30 \times 6.4 = 8.3$ ft. From Fig. 1 of the standard the basic wind pressure is $p = 40$ psf. From Table 3 of the standard the wind design pressures are

$$p_z = 30 \text{ psf, elevation} \quad 0 \text{ to } 30 \text{ ft} \quad \text{above ground}$$
$$40 \text{ psf,} \quad\quad\quad 30 \text{ to } 49 \text{ ft}$$
$$50 \text{ psf,} \quad\quad\quad 50 \text{ to } 99 \text{ ft}$$
$$60 \text{ psf,} \quad\quad\quad 100 \text{ to } 499 \text{ ft}$$

Wind loads in pounds per one foot of column height, $w = B \times D_e \times p_z$, are

$$w_1 = 0.60 \times 8.3 \times 30 = 150 \text{ lb/ft}$$
$$w_2 = 0.60 \times 8.3 \times 40 = 200 \text{ lb/ft}$$
$$w_3 = 0.60 \times 8.3 \times 50 = 250 \text{ lb/ft}$$
$$w_4 = 0.60 \times 8.3 \times 60 = 300 \text{ lb/ft.}$$

   b. Selected: 100 years recurrence interval; type-C exposure; damping factor $\beta = 0.01$; fundamental period of vibration $T = 1$ sec/cycle.
   From Fig. 2 of the standard: basic wind velocity is $V_{30} = 100$ mph; $q_{30} = 25.6$ psf; design pressures are $q_F = K_z G_F q_{30}$.
   From Fig. A2 of the standard:

$$K_{30} = 1.0, \quad K_{50} = 1.2, \quad K_{95} = 1.40, \quad K_{120} = 1.50.$$

Gust response factor is $G_F = 0.65 + 1.95\,(\sigma\sqrt{P}) = 0.65 + 1.95 \times 0.332 = 1.297$ for enclosed structures:

$$\sigma\sqrt{P} = 1.7\,[T(2h/3)]\,[0.785\,PF/\beta + S/(1 + 0.002d)]^{1/2} = 0.332$$

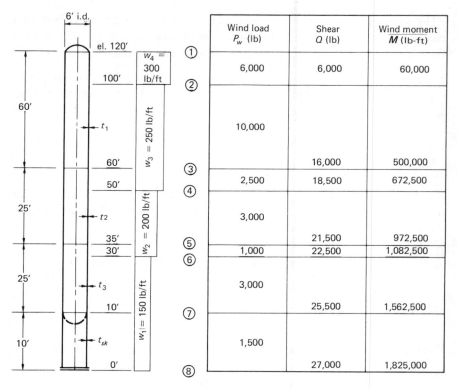

	Wind load P_w (lb)	Shear Q (lb)	Wind moment M (lb-ft)
①	6,000	6,000	60,000
②			
	10,000		
③		16,000	500,000
④	2,500	18,500	672,500
	3,000		
⑤		21,500	972,500
⑥	1,000	22,500	1,082,500
	3,000		
⑦		25,500	1,562,500
	1,500		
⑧		27,000	1,825,000

Fig. 1.4. Conditions and resulting wind loads for Example 1.1a.

where, from

Fig. A3. $T(2h/3) = T(80) = 0.149$

Fig. A4. $1.12(K)^{1/2} V_{30}/f = 1.12 \times 1 \times 100/1 = 112 \longrightarrow P = 0.092$

Fig. A5. $0.88\, fh/(V_{30}\sqrt{K_{120}}) = 0.88 \times 1.0 \times 120/100\sqrt{1.5} = 0.862$ and $h/c = h/d = 120/6.5 = 18.5 \longrightarrow F = 0.1$

Fig. A6. structure size factor: $S = 1.0$ for $h = 120$.

$$q_F = 25.6 \times 1.297 \times K_z = 33.2\, K_z,$$

therefore

$$q_{F30} = 33.2 \times 1 \doteq 34 \text{ psf}$$

$$q_{F50} = 33.2 \times 1.2 \doteq 40 \text{ psf}$$

$$q_{F95} = 33.2 \times 1.4 \doteq 47 \text{ psf}$$

Wind loads in pounds per foot of column height, $w = C_f \times D_e \times q_F$, are

$$w_1 = w_{30} = 0.66 \times 8.3 \times 34 = 186 \text{ lb/ft}$$

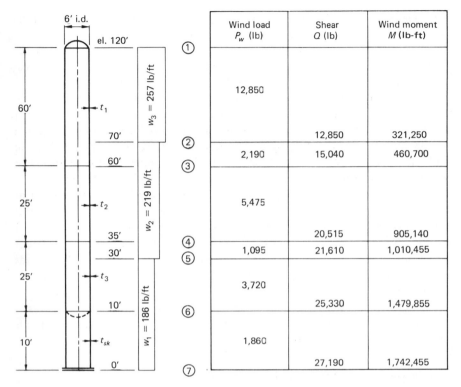

Fig. 1.5. Conditions and resulting wind loads for Example 1.1b.

$$w_2 = w_{50} = 0.66 \times 8.3 \times 40 = 219 \text{ lb/ft}$$

$$w_3 = w_{95} = 0.66 \times 8.3 \times 47 = 257 \text{ lb/ft.}$$

The method of determining wind loads on vessels of two or more diameters is the same as for a vessel of a uniform diameter. When the conical transition section is no more than 10 percent of the total height, cylindrical sections can be assumed to extend to the mid-height of the conical section. Otherwise, the transition section should be considered as a separate section.

1.6. EARTHQUAKE LOADS

General Considerations

Seismic forces on a vessel result from a sudden erratic vibratory motion of the ground on which the vessel is supported and the vessel response to this motion. The principal factors in the damage to structures are the intensity and the duration of the earthquake motion. The forces and stresses in structures during an earthquake are transient, dynamic in nature, and complex. An accurate analysis is generally beyond the kind of effort that can be afforded in a design office.

To simplify the design procedure the vertical component of the earthquake motion is usually neglected on the assumption that the ordinary structures possess enough excess strength in the vertical direction to be earthquake resistant.

The horizontal earthquake forces acting on the vessel are reduced to the *equivalent static forces*. Earthquake-resistant design is largely empirical, based on seismic coefficients derived from the performance of structures subjected in the past to severe earthquakes. The fundamental requirement set forth in building codes is that the structures in seismic risk zones must be designed to withstand a certain minimum horizontal shear force applied at the base of the vessel in any direction. Having assigned a minimum value to the base shear based on the past experience, the problem which arises is how to resolve this shear into equivalent static forces throughout the height of the vessel in order to determine the shear and the bending moments in the structure at different elevations as well as the overturning moment at the base. The result depends in large part on the dynamic response of the structure, which may be assumed either *rigid* or *flexible*.

For design purposes it is sufficient and conservative to assume that the vessel is fixed at the top of its foundation; no provision is usually made for any effects of the soil–structure interaction.

Seismic Design of a Rigid Cylindrical Vessel

The structure and its foundation are assumed to be rigid and the assumed earthquake horizontal acceleration of the ground a is transmitted directly into the vessel. The term *rigid* is used here in the sense of having no deformations.

Each section of the vessel will be acted upon by a horizontal inertial force equal to its mass and multiplied by the horizontal acceleration a of the quake movement,

$$\Delta P = \Delta W(a/g),$$

acting at the center of the gravity of the section. The overturning moment at an arbitrary elevation is equal to ΔP times the distance of the center of gravity of the vessel section above the section plane (see Fig. 1.6). Their resultant P_e is assumed to act at the center of gravity of the entire vessel and is given by the equation

$$P_e = Ma = (a/g)\, W = cW$$

where

    g = gravitational acceleration
    W = operating vessel weight during the earthquake

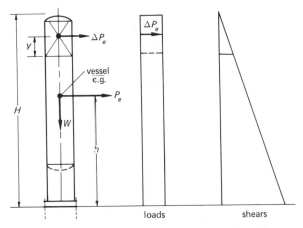

Fig. 1.6. Earthquake loads and shears for a rigid column of a uniform cross section and weight.

Table 1.2. Values of the Coefficient c.

ITEM	ZONE 0	ZONE 1	ZONE 2	ZONE 3
Vessel	0	0.05	0.10	0.20
Equipment attached to vessel	0	0.25	0.50	1.00

Source: Ref. 4.

$c = a/g$, an empirical seismic coefficient, depending on the seismic zone where the vessel is located.

The usual value of c, when used, is given in Table 1.2.

The overturning moment at the base M_b is equal to P_e times the elevation h of the center of gravity of the vessel above the vessel base:

$$M_b = P_e \times h.$$

The simple rigid-structure approach was used in early building codes. For a short heavy vessel or a horizontal drum on two supports this design procedure is easy to apply and probably justified. However, it cannot be reasonably applied to tall, slender process columns, regardless of their dynamic properties.

Seismic Design of Flexible Tall Cylindrical Vessels

The sudden erratic shift during an earthquake of the foundation under a flexible tall cylindrical vessel relative to its center of gravity causes the vessel to deflect, since the inertia of the vessel mass restrains the vessel from moving simultane-

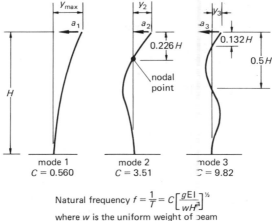

Natural frequency $f = \frac{1}{T} = C\left[\frac{gEI}{wH^4}\right]^{\frac{1}{2}}$

where w is the uniform weight of beam
per unit length.

Fig. 1.7. Sketch of mode shapes for a cantilever beam of a uniform cross section and weight.

ously with its foundation. The vibration initiated by the induced elastic deflection is then gradually reduced by damping or partial yielding in the vessel.

From experience and theoretical studies it can be assumed that a structure with a longer first period of vibration T and higher damping will be subjected to less total damage than a structure with a shorter T and smaller damping capacity, provided that it has the strength to withstand the sustained deflections.

A tall, slender vessel represents a distributed system of multiple degrees of freedom and will be excited into a transverse vibratory motion, consisting of relatively simple deflection curves called modes with the first fundamental mode predominant. Each mode has a unique period of vibration. The first three modes for a cantilever beam of uniform cross section and weight are illustrated in Fig. 1.7. Since the vessel will try to vibrate with a combination of natural frequencies, with the first frequency predominant, the resultant motion can become complex and determined by superposition. The force on a vessel section caused by vibration is then equal to the product of its mass and the vector sum of the accelerations associated with each mode. Each point under vibration can experience a maximum dynamic shear. Obviously, an involved, detailed mathematical analysis, based on still incomplete field data derived from observations of the past earthquakes would not be justified in the practical design.

For practical design purposes the building codes [3, 5]* require all free-standing structures in seismic zones to be designed and constructed to withstand the minimum lateral force V applied at the base in any horizontal direction and

*Numbers in brackets refer to the references in Appendix A8.

equal to the product of the weight and empirical coefficients:

$$V = ZKCW,$$

where

W = normal operating weight most likely to exist at the time of a possible future earthquake;

Z = earthquake zone factor, a numerical coefficient dependant upon the seismic zone in which the structure is located and found from the seismic zoning maps in ref. 3 (see also Appendix A1); for location in zone 1, $Z = 0.25$; zone 2, $Z = 0.50$; zone 3, $Z = 1$;

K = structure coefficient, a numerical coefficient related to the inherent resistance of a structure type to the dynamic seismic forces; K is based on the earthquake performance record of the type of the structure; for cylindrical vessels supported at base the K value is taken as equal to 2; for other structures K may vary from 0.67 to 3.00;

C = flexibility factor, a numerical coefficient depending on the flexibility of the vessel and given by the equation $C = 0.05/T^{1/3}$, where T is the fundamental period of vibration of the vessel, in seconds, in the direction under consideration; for $T < 0.12$ s, the factor C is usually taken as 0.10; a great accuracy of T is not needed in computing the coefficient C in the above equation, since C is inversely proportional to the cube root of T, and hence does not change appreciably with small variations of T, and the assumed fixity at the base will tend to make the calculated T smaller than the true period (see Section 4.7 for the procedure used to compute the basic period of vibration of tall, slender, self-supporting process columns).

The building codes prescribe the distribution of the base shear V over the height of the structures in accord with the triangular distribution equation

$$F_x = (V - F_t)\, w_x h_x \Big/ \sum_{i=1}^{n} w_i h_i$$

where

$F_t = 0.004\, V(h_n/D_s)^2 \leqslant 0.15\, V$ is a portion of V assumed concentrated at the uppermost level h_n of the structure to approximate the influence of higher modes. For most towers $F_t = 0.15\, V$, since $h_n/D_s > 6.12$;

F_i, F_x = the lateral force applied at levels h_i, h_x, respectively;

w_i, w_x = that portion of W which is located at or is assigned to levels i, x, respectively;

D_s = the plan diameter of the vessel.

For check, the total lateral shear at the base is $V = F_t + \sum_{i=1}^{n} F_i$.

The force acting lateraly in any direction on any appendage connected to the vessel is given by the equation

$$F_p = ZC_p W_p,$$

where W_p is the weight of the appendage and C_p is taken equal to 0.2. Force F_p is applied at the center of gravity of the attached equipment.

Since the higher modal responses contribute mainly only to the base shear, but not to the overturning moments, the base moment M_b and the moments M_x at levels h_x above the base can be reduced by means of reduction coefficients J and J_x, and are given by the following equations:

$$M_b = J \left(F_t h_n + \sum_{i=1}^{n} F_i h_i \right), \quad \text{where} \quad 0.45 < J = 0.6/T^{2/3} \leqslant 1$$

$$M_x = J_x \left[F_t (h_n - h_x) + \sum_{i=1}^{n} F_i (h_i - h_x) \right], \quad \text{where} \quad J_x = J + (1 - J)(h_x/h_n)^3.$$

For structures where the total mass is predominantly concentrated at one level and/or it would seem reasonable to expect the structure to vibrate primarily in the fundamental mode, $J = 1.0$ is recommended.

Example 1.2. Compute the seismic loads and moments acting on a cylindrical vertical process column of two diameters, as shown in Fig. 1.8.

To determine the earthquake load distribution on a vessel with variable mass concentrations or on a vessel of two or more diameters the entire vessel is usually divided into n sections; not more than 10 sections are usually required, the exact number depending on the weight distribution. The weight of each section w_i is assumed to be concentrated at the center of gravity of each section and the seismic loads F_i and the moments M_x are computed.

From Fig. 1.8 the total lateral force at the base V equals

$$V = ZKCW = 1 \times 2 \times 0.05 \times 85 = 8.5 \text{ kips}$$

where

$K = 2$
$W = 85$ kips, including steel shell, trays, operating liquid, insulation, etc.
$Z = 1$ for zone 3
$C = 0.05/T^{1/3} = 0.05/1^{1/3} = 0.05.$

In computing T it would seem to be conservative to assume that the period T for column 5-ft i.d. $\times$ 100 ft high with uniformly distributed weight $w = W/H$

will be shorter than that of the stepped-down column in Fig. 1.8. Using the formula for a uniform-diameter cantilever beam,

$$T = (2.70/10^5)(H/D)^2(wD/t)^{1/2} = (2.70/10^5)(100/5)^2(850 \times 5/0.5)^{1/2} = 1 \text{ sec.}$$

(For more accurate computation of T see the Example 4.4 in Section 4.7.) Force $F_t = 0.15 V = 1.3$ kips, and from Fig. 1.8,

$$F_i = (V - F_t) w_i h_i / \Sigma w_i h_i = 7.2 w_i h_i / 4630 \doteq 0.0016 w_i h_i.$$

The transverse design shear V_x at h_x elevations is equal to the sum of all lateral forces F_i above the section elevation h_x. Taking $J = 1.0$ the incremental moments at particular section planes are

$$\Delta M_x = F_t(H - h_x) + \Sigma F_i(h_i - h_x) \quad \text{or}$$

$$\Delta M_2 = 1.3 \times 17 + 3 \times 17/2 = 47.6 \text{ kips-ft}$$

$$\Delta M_3 = 4.3 \times 17 + 2.4 \times 17/2 = 93.5 \text{ kips-ft} \quad \text{etc.}$$

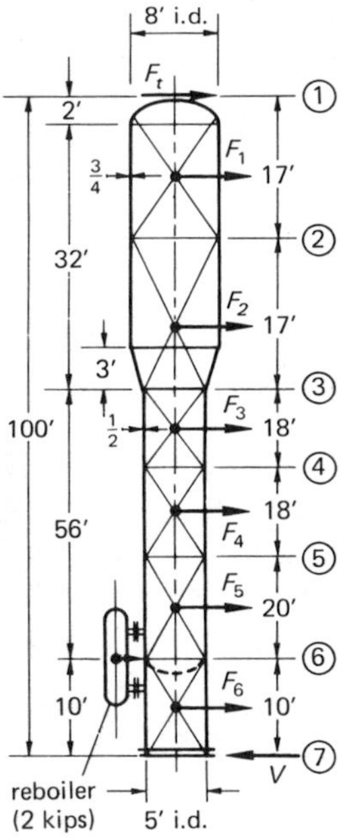

w_i (kips)	h_i (ft)	$w_i h_i$ (kips-ft)	F_i (kips)	V_x (kips)	ΔM_x (kips-ft)	M_x (kips-ft)	M_e (kips-ft)	h_x (ft)
100			1.3					
20	91.5	1830	3					
				4.3	47.6	47.6		83.0
20	74.5	1490	2.4					
				6.7	93.5	141.1		66.0
10	57.0	570	0.9					
				7.6	128.7	269.8		48.0
10	39.0	390	0.65					
				8.25	143	412.8		30.0
15	20.0	300	0.50					
				8.75	170	582.8		10.0
10	5.0	50	0.08					
				8.84	88	670.8	4	0.0
85 $\Sigma w_i = W$		4630 $\Sigma w_i h_i$		ΣV_x				

Fig. 1.8. Conditions and the resulting earthquake loads for Example 1.2.

The total moments acting at particular section planes are

$$M_2 = \Delta M_2 = 47.6 \text{ kips-ft}$$

$$M_3 = \Delta M_2 + \Delta M_3 = 141.1 \text{ kips-ft, etc.,}$$

as shown in Fig. 1.8. Total overturning moment at base,

$$M_7 = 670.8 + 4 = 674,800 \text{ lb-ft,}$$

is used to size the skirt base and the anchor bolts. Moments at other critical sections such as M_3 at section $h_x = 66$ ft or M_6 at section $h_x = 10$ ft are used for checking the stresses in the shell and the skirt-to-shell weld.

Example 1.3. Determine shear forces and moments due to seismic loads acting on cylindrical vertical column of a constant diameter and uniformly distributed weight w (lb per ft), as shown in Fig. 1.9.

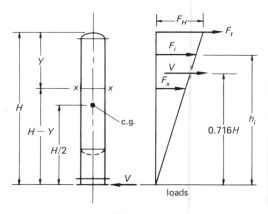

Fig. 1.9.

From the equation for the force $F_i = (V - F_t)\,w_i h_i / \Sigma w_i h_i$, with h_i the only independent variable, it can be seen that the distribution of F_i throughout the vessel height is triangular. If the total lateral force at base is $V = ZKCW$, at any horizontal plane at elevation h_x above the ground at section x-x, the shear force is

$$\begin{aligned}
V_x &= Y(F_H + F_x)/2 + F_t \\
&= \tfrac{1}{2}[(V - F_t)\,wH/(wH^2/2) + (V - F_t)\,w(H - Y)/(wH^2/2)]\,Y + F_t \\
&= (V - F_t)(2HY - Y^2)/H^2 + F_t
\end{aligned}$$

and the moment is

$$M_x = J_x[F_t Y + (V - F_t)(3HY^2 - Y^3)/3H^2]$$

At the base $Y = H$, the shear is

$$V_b = V - F_t + F_t = V$$

and the moment is

$$M_b = J[F_t H + (V - F_t)(2H/3)].$$

For $F_t = 0.15 V$ and $J = 1$,

$$M_b = 0.716 VH.$$

Design Considerations

It is only reasonable to assume that during a severe earthquake exceeding the assumed design value some part of the structure will yield and absorb the dynamic energy, preventing a major failure. The anchor bolts are the most logical structural part to prevent, through partial yielding any other damage to the vessel, such as buckling of the supporting skirt or shell. If the anchor bolts are to perform this function they should be long and resilient enough to yield under an extreme overload, and they should be firmly attached to the vessel, preferably through a full ring stiffener, as shown in Fig. 4.3, Type A.

Most codes allow an increase by one-third in the allowable stresses under earthquake (temporary) conditions. The ASME Pressure Vessel Code, Section VIII, Division 1 does not allow any increase in allowable stresses for the pressure parts, while Division 2 allows an increase of 20 percent in allowable stresses for the pressure parts and structural parts. At this point it would seem important to realize that the codes present only the minimum requirements, which should be increased by the designer according to his judgment after a careful assessment of all design conditions.

1.7. PIPING LOADS

In addition to the wind loads the piping loads acting on the vessel should be evaluated. They consist of the weight of the pipe sections supported by nozzles into the vessel shell and of the loads due to the thermal expansion of the pipes. The thermal expansion loads have to be estimated at the time of the vessel design. It can be assumed that the total sum of the piping reactions of all side nozzles will have a small effect on the entire vessel and can be disregarded. The thermal thrust at the top nozzle can be considerable, and in such case should be added to the other loadings acting on the vessel. The expansion pipe moment M_p will depend on the size of the nozzle, that is on the size of the process pipe.

Table 1.3.

	NOMINAL SIZE OF TOP NOZZLE (in.)	M_p (lb-ft)
M_p	4	2110
	6	4450
	8	7900
	10	13000
	12	19500
	14	24500
	16	34300
	18	46300
	20	61000
	24	98400

The everage M_p can be estimated as approximately equal to $M_p = 60D^3$ in lb-in., where D is the outside diameter of the pipe connected to the nozzle, increased by 3 in. (see ref. 7). Table 1.3 gives values of M_p computed by the above equation in lb-ft and could be added to the computed wind load. The actual moment and the required local reinforcement of the top shell head may not be as large.

1.8. COMBINATIONS OF THE DESIGN LOADS

Many combinations of loads considered in the design of pressure vessels may be possible, but highly improbable; therefore it is consistent with good engineering practice to select only certain sets of design loads, which can most probably occur simultaneously, as the design conditions for pressure vessels. If a more severe loading combination does occur, the built-in safety factor is usually large enough to allow only a permanent deformation of some structural member, without crippling damage to the vessel itself.

It is standard engineering practice that all vessels and their supports must be designed and constructed to resist the effects of the following combinations of design loads without exceeding the design limit stresses. (In all combinations wind and earthquake loads need not be assumed to occur simultaneously, and when a vessel is designed for both wind and earthquake, only the one which produces the greater stresses need be considered.)

1. *Erection* (*empty*) *design condition* includes the erection (empty) dead load of the vessel with full effects of wind or earthquake.

2. *Operating design condition* includes the design pressure plus any static liquid head, the operating dead load of the vessel itself, the wind or earthquake loads, and any other applicable operating effects such as vibration, impact and thermal loads.

3. *Test design condition for a shop hydrotest*, when the vessel is tested in a horizontal position, includes only the hydrotest pressure plus the shop test weight of the vessel. For a *field test* performed on location, the design condition includes the test pressure plus the static head of the test liquid, and the field test dead load of the vessel. Wind or earthquake loads need not be considered. All insulation or internal refractory are removed.

4. *Short-time (overload) design condition* includes the operating design condition plus any effects of a short-time overload, emergency, startup, or shutdown operations, which may result in increased design loads. At startup, the vessel is assumed to be cold and connecting pipelines hot. Wind or earthquake need not be considered.

The maximum stresses as computed from the above design conditions cannot exceed the design limit stresses; see Table 2.1 or AD-150.1 and Fig. 2.2, in Chapter 2 of this book.

2
Stress Categories and Design Limit Stresses

2.1. INTRODUCTION

After the design loads are determined and the maximum stresses due to the design loads are computed, the designer must qualitatively evaluate the individual stresses by type, since not all types of stresses or their combinations require the same safety factors in protection against failure.

For instance, when a pressure part is loaded to and beyond the yield point by a mechanical (static) force, such as internal pressure or weight, the yielding will continue until the part breaks, unless strain hardening or stress distribution takes place. In vessel design, stresses caused by such loads are called *primary* and their main characteristic is that they are not self limiting, i.e., they are not reduced in magnitude by the deformation they produce.

On the other hand, if a member is subjected to stresses attributable to a thermal expansion load, such as bending stresses in shell at a nozzle connection under thermal expansion of the piping, a slight, permanent, local deformation in the shell wall will produce relaxation in the expansion forces causing the stress. The stresses due to such forces are called *secondary* and are self limiting or self equilibrating.

The practical difference between primary and secondary loads and stresses is obvious; the criteria used to evaluate the safety of primary stresses should not be applied to the calculated values of stresses produced by self-limiting loads. Some stresses produced by static loads, such as the bending stresses at a gross structural discontinuity of a vessel shell under internal pressure, have the same self-limiting properties as thermal stresses and can be treated similarly.

Stresses from the dynamic (impact) loads are much higher in intensity than stresses from static loads of the same magnitude. A load is dynamic if the time of its application is smaller than the largest natural period of vibration of the body.

A structure may be subjected only rarely to the maximum wind or seismic load for which it has been designed. Therefore, an increase in allowable stresses is permitted for such temporary loads in some codes. Fatigue caused by periodic variation of mechanical or thermal loads over operating cycles has become an increasingly important consideration in the design of pressure vessels as it has become apparent that the majority of fractures are fatigue rather than static-loading failures. If the number of the operating cycles is larger than several thousand, fatigue analysis should be considered. Here the allowable stress and the stress range must be related to the number of loading cycles anticipated during the service life of the equipment. Fatigue failures usually occur in the zone of the maximum stress concentration.

The designer must not be content to understand the properties of the construction materials to be used in the vessel. He must also consider in detail the nature of the loads acting on the vessel (mechanical, thermal, cyclic, dynamic, static, temporary). Knowledge of these loads and the resulting stresses, obtained analytically and as accurately as possible, is essential to proper vessel design.

Suitable precautions expressed in the design safety factors are the responsibility of the design engineer, guided by the needs and specifications of the client. The chief requirement for the acceptibility of a Code-designed vessel is that the calculated stress levels shall not exceed Code allowable stress limits or, in their absence, stress limits based on the current good engineering practice.

2.2. ALLOWABLE STRESS RANGE FOR SELF-LIMITING LOADS

The most important self-limiting stresses in the design of pressure vessels are the stresses produced by thermal expansion and by internal pressure at shell structure discontinuities.

In the study of self-limiting stresses, fictitious elastic stress calculated as twice the yield stress has a very special meaning. It specifies the dividing line between the low cycle loads that, when successively applied, allow the structure to "shake down" to an elastic response, and loads that produce a plastic deformation every time they are applied. This can be illustrated in an idealized stress-strain diagram as shown in Fig. 2.1. Material is assumed to behave in elastic—perfectly plastic manner. Due to an applied thermal expansion load of the attached piping on a nozzle an elastic deformation occurs at some point in the vessel shell from O to A in Fig. 2.1 and a plastic irreversible deformation from A to B. At point B the thermal load is sufficiently reduced by the plastic deformation to be in equilibrium with the internal resisting stresses in the shell. When the nozzle and piping return to their original position, the stresses recover along line BCD. The elastic portion from C to D represents prestressing in compression of the permanently strained shell fibers in such a way that the next operating cycle from D to B lies entirely in the elastic range. Stress $\sigma_2 = 2\sigma_y$ represents the limit or the maximum stress range for elastic shakedown to be possible. At high operating temperatures the induced hot stress tends to diminish, but the sum of hot and cold stresses remains constant and is referred to as stress range.

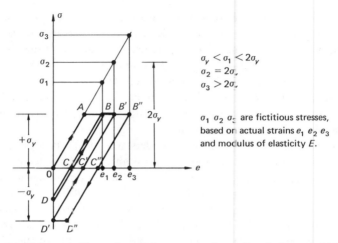

Fig. 2.1 Schematic illustration of the stress–strain relation during a shakedown.

The *shakedown load*, sometimes called the *stabilizing load*, is then the maximum self-limiting load that, when applied to a vessel, will on removal leave such built-in moments of the internal residual stresses that any subsequent application of the same or a smaller load will cause only elastic stresses in the vessel. If the plastic deformation from A to B'' is too large, then when the load is removed and the structure returns to the original state, a compressive yield from D'' to D' is introduced. At the next operating cycle the stress is in the plastic range again. Repeated plastic strains are particularly objectionable, since they lead to a failure in short time.

While the failure stress for the direct membrane stress σ due to a mechanical load is equal to the yield stress σ_y, and when the failure stress for the bending stress σ_b is $1.5\,\sigma_y$, only a limited, one-time, permanent deformation from points A to B occurs from self-limiting loads up to the computed stress σ equal to $2\sigma_y$.

From the above discussion it can be concluded that different stress limits can be set for maximum calculated stresses caused by different types of loads. In this way we can achieve more economical, but still safe design.

2.3. GENERAL DESIGN CRITERIA OF THE ASME PRESSURE VESSEL CODE, SECTION VIII, DIVISION 1

While Division 1 of the ASME Pressure Vessel Code, Section VIII provides the necessary formulas to compute the required thicknesses and the corresponding membrane stresses of the basic vessel components due to internal and external pressures, it leaves it up to the designer to use analytical procedures for computing the stresses due to other loads.

Paragraph UG-23c states that the wall thickness of a vessel computed by the Code rules shall be determined so that the maximum direct membrane stress, due to any combination of loadings as listed in UG-22 (internal or external pressure, wind loads or seismic loads, reaction from supports, the effect of temperature gradients, impact loads) that are expected to occur simultaneously during normal operation of a vessel, does not exceed the maximum allowable metal Code stress values permitted at the operating temperature. The direct membrane stress can be here defined as a normal stress uniformly distributed across the shell thickness of the section under consideration.

The above requirement implies the use of the maximum-stress theory of failure, on the assumption that for thin-shell pressure vessels the radial compressive stress σ_r due to the design pressure can be disregarded, and that the more accurate maximum-shear theory gives approximately the same results. The localized discontinuity stresses (membrane plus bending) are taken into account by the Code low allowable stresses and the Code approved design rules for details such as cone–cylinder junctions. Otherwise, detailed analytical stress analyses of secondary or fatigue stresses are not required and no design stress limits are imposed for them. Design limit stresses are not included for thermal expansion bending stresses and discontinuity bending stresses.

However, it is a general practice to provide detailed stress analysis for the vessel components outside the Code approved details using either the maximum-stress or the maximum-shear theory of failure, and to select allowable stresses for design conditions other than normal operations or for computed stresses other than direct membrane or direct membrane plus primary bending Code stresses.

Table 2.1 gives recommended design allowable stresses for stresses due to various loads not included in UG-23c. As noted above, design limit stresses for secondary stresses in combination with pressure stresses are not specified in the Code. However, paragraph UA-5e allows higher allowable stresses for stresses at cone–cylinder junctions; this would indicate that different stress levels for different stress categories are acceptable. In conclusion, it would be well to remember, as previously pointed out, that the Code requirements represent only the minimum, and the designer should feel free to apply stricter design limits when necessary according to his judgment and experience.

Code allowable stresses are generally used also for the design of *important nonpressure parts*, such as support skirts for tall columns, supports for functionally important vessel internals, and also for the welds attaching such parts with appropriate joint efficiencies. The allowable stresses for the less important *nonpressure replaceable structural parts* may be higher than the Code stresses and can be conveniently taken from ref. 9. Reference 12 serves as a useful guide for selection of the allowable stresses of *nonpressure parts at high temperatures*.

Table 2.1. Design Limit Stresses for Pressure Parts.

DESIGN CONDITION	COMBINATION OF DESIGN LOADS						MAXIMUM ALLOWABLE STRESS	NOTE
	DESIGN PRESSURE	VESSEL WEIGHT	WIND OR SEISMIC LOAD	TEMPORARY OVERLOAD	LOCAL MECHANICAL LOAD	THERMAL EXPANSION LOAD		
1. Field erection		×	×				$1.2S_a$	1
2. Operating	×	×	×				S_a	2, 5
3. Operating with thermal expansion	×	×	×			×	$1.25(S_a + S_{atm})$	2, 3, 5
4. Short-term mechanical overload	×	×		×			$1.33S_a < S_y$	2
5. Short-term mechanical overload with thermal expansion	×	×		×		×	$1.50(S_a + S_{atm})$	2
6. Operating with local mechanical load (bending)	×	×	×		×		$2S_a < S_y$	2, 5
7. Hydrotest	test pressure		×				$S_{atm} \times$ test factor	4

S_a = basic Code allowable stress value at design metal temperature
S_{atm} = basic Code allowable stress value at atmospheric metal temperature

Notes

1. Computed stresses are based on the full uncorroded thickness.

2. Computed stresses are based on the corroded thickness.

3. The criterion used here is the approach used in the ANSI B31.3 Petroleum Refinery Piping, that is, to limit the maximum calculated principal stress to the allowable stress range rather than using the shear theory of failure. The value $1.25(S_a + S_{atm})$ provides a safety margin against the possibility of fatigue due to localized stresses and other stress conditions (see Fig. 2.1). Obviously, stresses must be computed both with and without thermal expansion, since allowable stresses are much smaller for conditions without thermal expansion. As an additional safety precaution, the computed stresses are usually based on the modulus of elasticity E at room temperature [8].

4. Computed stresses are based on test thickness at test temperature. Since water pressure is a short-term condition, the allowable stresses for structural parts such as supports are frequently increased by a factor of 1.2. The upper limit of stress in the vessel shell during a hydrotest of pressure parts is not specified by the Code. However, it is a good engineering practice to limit the maximum membrane stress in any part of the vessel during a hydrotest to 80 percent of the yield strength. Thermal expansion stresses and local mechanical stresses will be absent and need not be considered.

5. This allowable stress limit is for statically stressed vessels. This means that the number of stress cycles applied during the life of vessel does not exceed several thousands. For cyclic conditions these allowable stresses would have to be substantially reduced and preferably a fatigue analysis made. Usually, in service the operating loads are raised gradually up to their maximum values and maintained for some time and they are not reapplied often enough to make a fatigue analysis necessary.

2.4. GENERAL DESIGN CRITERIA OF THE ASME PRESSURE VESSEL CODE, SECTION VIII, DIVISION 2

Higher basic allowable stresses than in the Code Division 1 are permitted to achieve material savings in vessel construction. Also increased stress limits for various load combinations are allowed by using the factor k in Table AD-150.1, not allowed in Division 1. To preserve the high degree of safety, strict design, fabrication, and quality-control requirements are imposed.

The most important points can be summarized as follows.

1. Specification of the design conditions, including all sufficient data pertaining to the method of support, type of service (static or cyclic), and type of corrosion, is the responsibility of the user. The report must be certified by a registered professional engineer.

2. The structural soundness of the vessel becomes the responsibility of the manufacturer, who is required to prepare all design computations proving that the design as shown on the drawings complies with the requirements of the Division 2. Again, a registered professional engineer experienced in the design of pressure vessels has to certify the design report. Stress classification and a detailed stress analysis are required. Maximum-shear failure theory is used in preference to maximum-distortion-energy theory not only for its ease of application, but also for its directional applicability to fatigue stress analysis. Specific design details for vessel parts under pressure are provided, as well as the rules and guidance for analytical treatment of some types of loadings. A set of conditions is established (AD-160) under which a detailed fatigue analysis is required. Evaluation of thermal stresses is also required.

3. Strict quality control must be maintained by the manufacturer. Additional tests (ultrasonic, impact, weld inspection) are imposed which are not required in Division 1.

Stress Categories

One of the design requirements of Division 2 is an accurate classification of stresses according to the loads that cause them, their distribution, and their location. Division 2 establishes different allowable stress limits (stress intensities) for different stress categories.

Basically, the stresses as they occur in vessel shells (see Fig. 2.2), are divided into three distinct categories, *primary*, *secondary*, and *peak*.

1. *Primary stress* is produced by steady mechanical loads, excluding discontinuity stresses or stress concentrations. Its main characteristic is that it is not self limiting. Primary stress is divided into two subcategories; *general* and *local*.

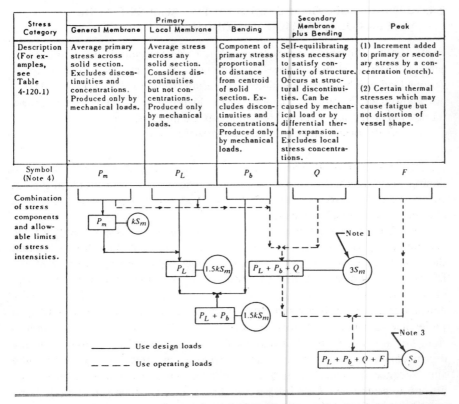

Stress Category	Primary			Secondary Membrane plus Bending	Peak
	General Membrane	Local Membrane	Bending		
Description (For examples, see Table 4-120.1)	Average primary stress across solid section. Excludes discontinuities and concentrations. Produced only by mechanical loads.	Average stress across any solid section. Considers discontinuities but not concentrations. Produced only by mechanical loads.	Component of primary stress proportional to distance from centroid of solid section. Excludes discontinuities and concentrations. Produced only by mechanical loads.	Self-equilibrating stress necessary to satisfy continuity of structure. Occurs at structural discontinuities. Can be caused by mechanical load or by differential thermal expansion. Excludes local stress concentrations.	(1) Increment added to primary or secondary stress by a concentration (notch). (2) Certain thermal stresses which may cause fatigue but not distortion of vessel shape.
Symbol (Note 4)	P_m	P_L	P_b	Q	F

Combination of stress components and allowable limits of stress intensities.

P_m — kS_m

P_L — $1.5kS_m$

$P_L + P_b + Q$ — $3S_m$ — Note 1

$P_L + P_b$ — $1.5kS_m$

$P_L + P_b + Q + F$ — S_a — Note 3

——— Use design loads

– – – – Use operating loads

NOTE.1 — This limitation applies to the range of stress intensity. When the secondary stress is due to a temperature excursion at the point at which the stresses are being analyzed, the value of S_m shall be taken as the average of the S_m values tabulated in Part AM for the highest and the lowest temperature of the metal during the transient. When part or all of the secondary stress is due to mechanical load, the value of S_m shall be taken as the S_m value for the highest temperature of the metal during the transient.

NOTE 2 — The stresses in Category Q are those parts of the total stress which are produced by thermal gradients, structural discontinuities, etc., and do not include primary stresses which may also exist at the same point. It should be noted, however, that a detailed stress analysis frequently gives the combination of primary and secondary stresses directly and, when appropriate, this calculated value represents the total of P_m (or P_L) + P_b + Q and not Q alone. Similarly, if the stress in Category F is produced by a stress concentration, the quantity F is the additional stress produced by the notch, over and above the nominal stress. For example, if a plate has a nominal stress intensity, S, and has a notch with a stress concentration factor, K, then $P_m = S$, $P_b = 0$, $Q = 0$, $F = P_m (K-1)$ and the peak stress intensity equals $P_m + P_m (K-1) = KP_m$.

NOTE 3 — S_a is obtained from the fatigue curves, Figs. 5-110.1, 5-110.2 and 5-110.3. The allowable stress intensity for the full range of fluctuation is 2 S_a.

NOTE 4 — The symbols P_m, P_L, P_b, Q, and F do not represent single quantities, but rather sets of six quantities representing the six stress components σ_t, σ_l, σ_r, τ_{tb}, τ_{lr}, and τ_{rt}.

NOTE 5 — The k factors are given in Table AD-150.1.

Fig. 2.2. Stress categories and limits of stress intensity. (Reproduced from the ASME Boiler and Pressure Vessel Code, Section VIII, Division 2 (1977 edition) by permission of the American Society of Mechanical Engineers.)

(a) *General primary stress*, is imposed on the vessel by the equilibration of external and internal mechanical forces. Any yielding through the entire shell thickness will not distribute the stress, but will result in gross distortions, often carried to failure. General primary stress is divided into primary membrane stress and primary bending stress; the limit design method shows that a higher stress limit can be applied to the primary bending stress than to the primary membrane stress. Typical examples of general primary membrane stress in the vessel wall are: stress due to internal or external pressure and stress due to vessel weight or external moments caused by wind or seismic forces. A typical example of primary bending stress is the bending stress due to pressure in flat heads.

(b) *Local primary stress* is produced by the design pressure alone or by other mechanical loads. It has some self-limiting characteristics. If the local primary stress exceeds the yield point of the material, the load is distributed and carried by other parts of the vessel. However, such yielding could lead to excessive and unacceptable deformations, so it is necessary to assign a lower allowable stress limit to this type of stress than to secondary stresses. An important property of local primary stress is that the maximum stress remains localized and diminishes rapidly with distance from the point of load application. Local primary stress can be divided into direct membrane stress and bending stress. Both, however, have the same stress intensity limits. Typical examples of local primary stress are stresses at supports and local membrane stresses due to internal pressure at structural discontinuities.

2. The basic characteristic of *secondary stress* is that it is self-limiting. Minor yielding will reduce the forces causing excessive stresses. Secondary stress can be divided into membrane stress and bending stress, but both are controlled by the same limit stress intensities. Typical examples of secondary stress are thermal stresses and local bending stresses due to internal pressure at shell discontinuities.

3. *Peak stress* is the highest stress at some local point under consideration. In case of failure, peak stress does not generate any noticeable distortion, but it can be a source of fatigue cracks, stress-corrosion, and delayed fractures. Generally, the computation of the peak stresses is required only for vessels in cyclic service as defined by AD-160. Typical examples of peak stress are thermal stress in carbon steel plate with stainless steel integral cladding and stress concentrations due to local structural discontinuities such as a notch, a small-radius fillet, a hole, or an incomplete penetration weld.

Combination of Stress Intensities

According to the shear theory of failure used in Division 2, yielding in a member under loads begins if the maximum induced shear stress equals the yield shear stress developed in a test sample under simple tension. The maximum shear stress τ at the point under consideration equals one-half the largest algebraic difference between any two out of three principal stresses $(\sigma_1, \sigma_2, \sigma_3)$ at that point. It occurs on each of two planes inclined 45 degrees from these two principal stresses. For $\sigma_3 > \sigma_2 > \sigma_1$, maximum $\tau = (\sigma_3 - \sigma_1)/2$. Twice the maximum shear stress τ is by definition the *equivalent intensity of combined stress* or *stress intensity*.

The procedure for computation of the stress intensities can be summarized as follows.

1. At a point under investigation the designer selects an orthogonal set of coordinates: L, t, and r.

2. The stresses due to the design loads and moments are calculated and decomposed into orthogonal components σ_r, σ_L, and σ_t parallel to the coordinates.

3. According to the types of loads, the stresses or their components are classified and combined in the following way:

P_m = sum of all general primary membrane stress components
P_L = sum of all local primary membrane stress components
P_b = sum of all primary bending stress components
Q = sum of all secondary (membrane and bending) stress components
F = sum of all peak stress components.

P_m, P_L, P_b, Q, and F can represent a triaxial stress combined with shear, and as such they would be defined by six stress resultants. In this case the principal stresses $(\sigma_1, \sigma_2, \sigma_3)$ must first be evaluated for each category separately. However, the coordinates (L, r, t) can be and usually are chosen in such a way that the stresses σ_L, σ_r, σ_t are already the principal stresses of the particular stress category.

4. The maximum stress intensity S is calculated for the particular stress type (e.g., P_m) or combination of types (e.g., $P_L + P_b + Q$) and compared with basic Code stress intensity limits. Given principal stresses $\sigma_1, \sigma_2, \sigma_3$ in the category under consideration, call

$$S_{12} = \sigma_1 - \sigma_2, \qquad S_{23} = \sigma_2 - \sigma_3, \qquad S_{31} = \sigma_3 - \sigma_1.$$

Then S is the absolute value of the largest of these differences:

$$S = \max\left(|S_{12}|, |S_{23}|, |S_{31}|\right).$$

Basic Stress Intensity Limits

One of the requirements for a design to be acceptable is that the computed stress intensities S will not exceed the allowable stress intensity limits of Division 2. As shown in Fig. 2.2 there are five basic allowable stress intensity limits to be met.

1. *General primary membrane stress intensity.* The maximum stress intensity S as computed on basis of P_m stresses cannot exceed a stress intensity equal to

S_m times the factor k from Table AD-150.1 when applicable $(S_m \leqslant \frac{2}{3} S_y$ or $\leqslant \frac{1}{3} S_u)$. S_m is the basic allowable design stress value (in tension) for approved materials.

2. *Local membrane stress intensity.* The maximum allowable stress intensity for S derived from P_L stresses is $1.5 S_m$ times the factor k when applicable.

3. *Primary membrane (P_m or P_L) plus primary bending (P_b) stress intensity.* The maximum stress intensity S based on P_m or P_L stresses plus the bending stress P_b is limited to $1.5 S_m$ times the factor k when applicable.

4. *Primary plus secondary stress intensity.* The maximum stress intensity S as based on the primary or local membrane stresses plus the primary bending stress plus the secondary stress $(P_m$ or $P_L + P_b + Q)$ cannot exceed the value of $3 S_m$. All stresses may be computed under operating conditions, usually less severe than the design conditions $(3 S_m \leqslant 2 S_y)$.

5. *Peak stress intensity.* If fatigue analysis is required for cyclic operating conditions, the maximum stress intensity S must be computed from the combined primary, secondary, and peak stresses $(P_m$ or $P_L + P_b + Q + F)$ under operating conditions. The allowable value S_a for this peak stress intensity S is obtained by the methods of analysis for cyclic operations with the use of the fatigue curves.

Table AD-150.1. Stress Intensity k Factors for Various Load Combinations.

Condition	Load Combination (See AD-110)	k Factors	Calculated Stress Limit Basis
Design	A The design pressure, the dead load of the vessel, the contents of the vessel, the imposed load of the mechanical equipment, and external attachment loads	1.0	Based on the corroded thickness at design metal temperature
	B Condition A above plus wind load	1.2	Based on the corroded thickness at design metal temperature
	C Condition A above plus earthquake load	1.2	Based on the corroded thickness at design metal temperature
	(NOTE: The condition of structural instability or buckling must be considered)		
Operation	A The actual operating loading conditions. This is the basis of fatigue life evaluation	See AD-160 and Appendix 5	Based on corroded thickness at operating pressure and metal operating temperature
Test	A The required test pressure, the dead load of the vessel, the contents of the vessel, the imposed load of the mechanical equipment, and external attachment loads	1.25 for hydrostatic test and 1.15 for pneumatic test. See AD-151 for special limits.	Based on actual design values at test temperature

Reproduced from the ASME Boiler and Pressure Vessel Code, Section VIII, Division 2 (1977 edition), by permission of the American Society of Mechanical Engineers.

In addition to the above conditions the algebraic sum of all three principal stresses should not exceed $4S_m$: $\sigma_1 + \sigma_2 + \sigma_3 \leqslant 4S_m$.

The following definitions apply in the above discussion:

Membrane stress is a normal stress component (tension or compression) uniformly distributed across the thickness of the wall section.

Bending stress is a normal stress component linearly distributed across the thickness about the neutral axis of the wall section.

Shear stress is a stress component tangent to the plane of the section and usually assumed uniformly distributed across the section.

2.5. DESIGN REMARKS

It can safely be said that the savings in material for ordinary carbon steel or low-alloy steel vessels designed according to the rules of Division 2 will be more than offset by the additional engineering and fabrication costs.

Since Division 2 is not used as often and as long as Division 1, most vessel designers are more familiar with Division 1.

Geometrically simple important vessels with small external loadings such as spherical reactors could probably be designed with some savings in accordance with the rules of the Division 2.

3
Membrane Stress Analysis of Vessel Shell Components

3.1. INTRODUCTION

In structural analysis, all structures with shapes resembling curved plates, closed or open, are referred to as shells. Obviously, in pressure vessel design the shells are always closed. Most pressure vessels in industrial practice basically consist of few shapes: spherical or cylindrical with hemispherical, ellipsoidal, conical, toriconical, torispherical, or flat end closures. The shell components are welded together, sometimes bolted together by means of flanges, forming a shell with a common rotational axis.

Generally, the shell elements used are axisymmetrical surfaces of revolution, formed by rotation of a plane curve or a simple straight line, called a meridian or generator, about an axis of rotation in the plane of the meridian (see Fig. 3.1). The plane is called meridional plane and contains the principal meridional radius of curvature. Only such shells will be considered in all subsequent discussions.

For analysis, the geometry of such shells has to be specified using the form of the midwall surface, usually the two principal radii of curvature, and the wall thickness at every point. A point on a shell, e.g., the point a in Fig. 3.1, can be located by the angles θ, ϕ and the radius R. In engineering strength of materials a shell is treated as *thin* if the wall thickness is quite small in comparison with the other two dimensions and the ratio of the wall thickness t to the minimum principal radius of curvature is $R_t/t > 10$ or $R_L/t > 10$. This also means that the tensile, compressive, or shear stresses produced by the external loads in the shell wall can be assumed to be equally distributed over the wall thickness.

Further, most shells used in vessel construction are *nonshallow* thin (membrane) shells in the range $1/500 > t/R > 1/10$, whose important characteristic is that bending stresses due to concentrated external loads are of high intensity only in close proximity to the area where the loads are applied. The attenuation length or the decay length, the distance from the load where the stresses almost die out, is limited and quite short—e.g., for a cylinder, $(Rt)^{1/2}$.

The radial deformation ΔR of the shell under a load is assumed to be small

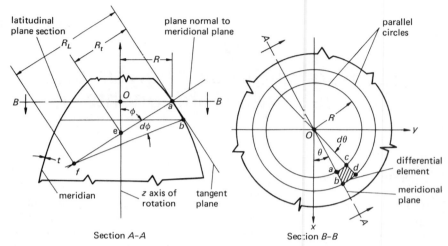

Section A-A   Section B-B

R_t = tangential (circumferential) radius of curvature of the surface, also R_θ.
R_L = longitudinal (meridional) radius of curvature of the surface, also R_ϕ.

Both radii of curvature R_L and R_t lie on the same line, but have different lengths (except for sphere, $R_L = R_t$).

$R = R_t \sin \phi$, radius of the parallel circle.
t = shell thickness.
ϕ = angle in the meridional plane section between the plane normal to the meridional plane and the axis of rotation.
θ = angle in latitudinal plane section between meridional plane and reference meridional plane (xz).

Area $abcd$, the differential shell element $R_L \, d\phi \times R_t \, d\theta$, is cut out by two meridians and two parallel circles, since it is convenient to investigate the stresses in the meridional and latitudinal planes. All radii are mean radii of shell wall in corroded condition.

Fig. 3.1. Definitions pertaining to shells of revolution.

with respect to the wall thickness ($<t/2$) and the maximum stresses remain below the proportional limit of the shell material.

If a general external (surface) load is acting on the shell, the loading on a shell element can be divided into three components; P_ϕ, P_θ, and P_R as shown in Fig. 3.2a. A thin, elastic shell element resists loads by means of internal (body) stress resultants and stress couples, acting at the cross sections of the differential element, as shown separately for clarity in Fig. 3.2b, c, and d. The surface forces act on the surface, outside or inside, while the body forces act over the volume of the element. Since the element must be in equilibrium, static equilibrium equations can be derived.

In general, there are 10 different internal stress resultants:

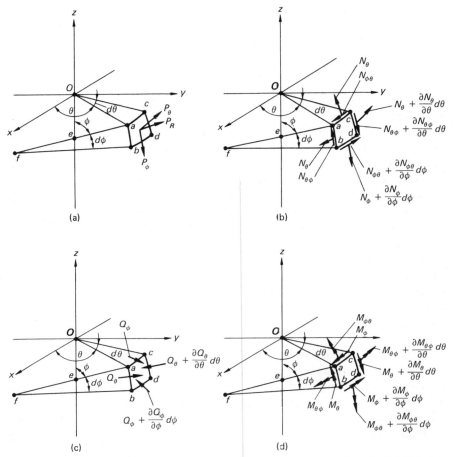

(a) General external load components P_R, P_ϕ, P_θ acting on a differential element of an axially symmetrical shell. For uniform internal pressure P, $P_\phi = P_\theta = 0$ and $P_R = P$.

(b) Shell plane membrane stress resultants N_ϕ, N_θ in tension or compression and $N_{\phi\theta}$, $N_{\theta\phi}$ in shear. Neglecting the trapezoidal shape of the element $N_{\phi\theta} = N_{\theta\phi}$. Membrane state of stress. For axisymmetrically loaded shells, $N_{\phi\theta} = N_{\theta\phi} = 0$ and $(\partial N_\theta/\partial\theta)\,d\theta = 0$.

(c) Transverse shear stress resultants Q_ϕ and Q_θ. For axisymmetrically loaded shells $Q_\theta = 0$.

(d) Bending stress resultant couples M_ϕ, M_θ and torque resultant couples $M_{\phi\theta}$, $M_{\theta\phi}$. Torque couples are neglected in bending shell theory. $M_{\theta\phi}$ and $M_{\phi\theta} = 0$ for axisymmetrically loaded shells.

Fig. 3.2. Stress resultants (tension, shear), bending and twisting stress couples at a differential element of an axisymmetrical shell under general load components P_R, P_ϕ, and P_θ.

$N_\phi, N_\theta, N_{\phi\theta}, N_{\theta\phi}$ = membrane forces acting in the plane of the shell surface, lb/in.

Q_ϕ, Q_θ = transverse shear, lb/in.

M_ϕ, M_θ = bending stress couples, lb-in./in.

$M_{\phi\theta}, M_{\theta\phi}$ = twisting stress couples, lb-in./in.

which must be in equilibrium with the external forces. Since there are only six equations of static equilibrium available for solution, the problem becomes four times statically indeterminate. From the geometry of the shell before and after deformation under the load, the direct and shear strains, changes in curvatures and twists can be established, and assuming a linear stress–strain relationship (Hooke's law) the additional required differential equations relating the stress resultants to the midshell surface displacements can be obtained. A rigorous solution of such a system of differential equations within the given boundary conditions is very difficult and can be accomplished only when applied to some special cases. Once the stress resultants are determined the stresses in the shell can be computed.

Fortunately, most vessel problems occurring in practice can be solved with satisfactory results using a simplified approach. The main reason for this is that under certain loading conditions which occur in practice with shells of revolution, some stress resultants are very small and can be disregarded or, because of axial symmetry, are equal to zero.

Membrane shell theory solves shell problems where the internal stresses are due only to membrane stress resultants N_ϕ, N_θ, and $N_{\phi\theta} = N_{\theta\phi}$ (see Fig. 3.2b). The shear stress resultants ($N_{\phi\theta}$, $N_{\theta\phi}$) for axisymmetrical loads such as internal pressure are equal to zero, which further simplifies the solution. The membrane stress resultants can be computed from basic static equilibrium equations and the resultant stresses in shell are:

longitudinal stress: $\sigma_L = \sigma_\phi = N_\phi/t$

tangential stress: $\sigma_t = \sigma_\theta = N_\theta/t$

Bending shell theory, in addition to membrane stresses, includes bending stress resultants (Fig. 3.2d) and transverse shear forces (Fig. 3.2c). Here the number of unknowns exceeds the number of the static equilibrium conditions and additional differential equations have to be derived from the deformation relations. Once the membrane stress resultants N_ϕ and N_θ and the resultant moments M_ϕ, M_θ are determined the stresses in shell are:

longitudinal stress $\sigma_L = \sigma_\phi = N_\phi/t \pm 6M_\phi/t^2$

tangential stress $\sigma_t = \sigma_\theta = N_\theta/t \pm 6M_\theta/t^2$

shear stress $\tau_\phi = Q_\phi/t.$

3.2. MEMBRANE STRESS ANALYSIS OF THIN-SHELL ELEMENTS

In most practical cases the loads on a vessel act in such a way that the reacting stress resultants shown in Fig. 3.2b will be predominant and the bending and transverse shear forces so small that they can be neglected. Here there are only three unknowns, N_ϕ, N_θ, and $N_{\phi\theta} = N_{\theta\phi}$, which can be determined from the equations for static equilibrium.

The main conditions for a membrane analysis to be valid can be summarized as follows.

1. All external loads must be applied in such a way that the internal stress reactions are produced in the plane of the shell only. Membrane stress analysis assumes that the basic shell resistance forces are tension, compression, and shear in the shell plane and that a thin shell cannot respond with bending or transverse shear forces. In practice, all thin shells can absorb some load in bending, but these bending stresses are considered secondary and are neglected. If under a concentrated load or edge loading conditions the bending stresses reach high values, a more detailed analysis has to be made and the shell locally reinforced if necessary.

2. Any boundary reactions, such as those at supports, must be located in the meridional tangent plane, otherwise transverse shear and bending stresses develop in the shell boundary region.

3. The shell including the boundary zone must be free to deflect under the action of the stress resultants. Any constraints cause bending and transverse shear stresses in the shell.

4. The change in meridional curve is slow and without cusps or sharp bends. Otherwise bending and transverse shear stresses will be included at such gross geometrical discontinuities.

5. The membrane stress resultants are assumed uniformly distributed across the wall thickness. This can be assumed if the ratio of the radius of curvature to the wall thickness is about $R/t \geqslant 10$ and the change in the wall thickness if any is very gradual.

6. The radial stress σ_r is small and can be neglected. A plane state of stress is assumed.

7. The middle surface of the entire shell is assumed to be continuous from one section of the shell component to another across any discontinuity. In practical vessel design two shell sections of different thicknesses are welded together to give a smooth inside contour. At the junction the lines of action of the meridional stress resultant N_ϕ are not collinear and this eccentricity introduces additional bending stresses.

8. The loadings are such that the shell deflections are small ($R \leqslant t/2$) and in the elastic range.

Membrane Stresses Produced by Internal Pressure

The most important case in the vessel design is a thin-shell surface of revolution, subjected to internal pressure of intensity P (measured in psi). The internal pressure is an axisymmetrical load; it can be a uniform gas pressure or a liquid pressure varying along the axis of rotation. In the latter case usually two calculations are performed to find the stresses due to equivalent gas pressure plus the stresses due to the liquid weight.

A freebody diagram of a shell element under uniform internal pressure is shown in Fig. 3.3. Due to uniform pressure and axisymmetry there are no shear stresses on the boundaries of the element $abcd$ ($N_{\phi\theta} = N_{\theta\phi} = 0$). The stresses σ_L and σ_t are the principal stresses and they remain constant across the element ($\partial N_\theta/\partial\theta) \, d\theta = 0$ and $(\partial N_\phi/\partial\phi) \, d\phi = 0$. The first equilibrium equation in the direction normal to the shell element (see Fig. 3.3) is

$$P[2R_t \sin(d\theta/2)] \, [2R_L \sin(d\phi/2)] = 2\sigma_t \, ds_1 \, t \sin(d\theta/2) + 2\sigma_L \, ds_2 \, t \sin(d\phi/2).$$

Substituting $\sin(d\theta/2) \doteq (ds_2/2)/R_t$ and $\sin(d\phi/2) \doteq (ds_1/2)/R_L$,

$$P \, ds_1 \, ds_2 = (\sigma_t/R_t) \, t \, ds_1 \, ds_2 + (\sigma_L/R_L) \, t \, ds_1 \, ds_2$$

or

$$P/t = (\sigma_t/R_t) + (\sigma_L/R_L).$$

This last equation (the first equilibrium equation) is of fundamental importance for stress analysis of axisymmetrical membrane shells subjected to loads sym-

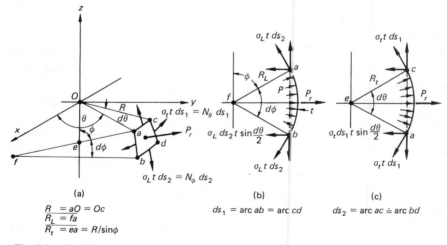

$$\frac{R}{R_L} = aO = Oc$$
$$R_L = fa$$
$$R_t = ea = R/\sin\phi$$

$ds_1 = $ arc $ab = $ arc cd

$ds_2 = $ arc $ac \doteq $ arc bd

Fig. 3.3.   Shell element of axisymmetrical shell subjected to uniform internal pressure P.

metrical with respect to the axis of rotation. The principal radius of curvature R_L can be positive (if it points toward the vessel axis), be negative (if it points away from the vessel axis), or become infinite (at an inflection point). In the above analysis the radial stress σ_r was assumed to be negligible (average $\sigma_r = P/2$).

The second equation of equilibrium required to solve for σ_t and σ_L can be obtained by summation of the forces and stresses in the direction of the axis of rotation. Since the shell is axisymmetrical the entire finite shell section can be used at once as follows:

$$2\pi R (t\sigma_L \sin \phi) = P\pi R^2$$

$$\sigma_L = PR/2t \sin \phi = PR_t/2t.$$

Substituting this value into the first equilibrium equation yields

$$\sigma_t = (PR_t/t)[1 - (R_t/2R_L)].$$

If $e_t = (1/E)(\sigma_t - \nu\sigma_L)$ is the unit elongation in the tangential direction, the radial growth ΔR can be derived as follows.

$$2\pi (R + \Delta R) = 2\pi R + 2\pi Re_t$$

$$\Delta R = Re_t = (R/E)(\sigma_t - \nu\sigma_L),$$

where E is the modulus of elasticity.

The third equation required for static equilibrium of a biaxial state of stress is automatically satisfied, since load and the resultant stresses in the tangential direction are defined as symmetrical with respect to the axis of rotation.

At this point it would seem important to point out that a change in the radius of curvature will introduce in the shell bending stresses that the membrane stress analysis assumes negligible. Taking a spherical shell as an example $R_L = R_t = R$ and stresses $\sigma = \sigma_t = \sigma_L = PR/2t$ due to the internal pressure, the unit shell elongation e_t is given by the following equation:

$$e_t = (\sigma/E)(1 - \nu) = (PR/2tE)(1 - \nu).$$

The change in curvature from R to $R' = (R + \Delta R) = R(1 + e_t)$ is given by

$$1/\Delta R = (1/R') - (1/R) = -(e_t/R)/(1 + e_t) \doteq -e_t/R = -(P/2tE)(1 - \nu).$$

For a thin spherical plate with equal curvatures in two perpendicular directions the unit edge bending moment causing the change $1/\Delta R$ in the curvature is given by $M = D(1 + \nu)/\Delta R$, where $D = Et^3/12(1 - \nu^2)$ is the flexural rigidity of the shell. Substituting for $1/\Delta R$ yields

$$M = -D(1 - \nu^2)(P/2tE) = -(Pt^2/24).$$

A negative moment will cause compressive stress in outside fibers.
   The bending stress is $\sigma_b = \pm 6M/t^2 = \pm P/4 \ll \sigma = PR/2t$ or

$$|\sigma/\sigma_b| = 2R/t.$$

Obviously the bending stress σ_b can be neglected, as was the radial stress σ_r.

Asymmetrically Loaded Membrane Shells of Revolution

If there are asymmetries in the load application, a shear stress resultant $N_{\phi\theta}$ will be induced in addition to two other membrane stress resultants N_ϕ and N_θ as in Fig. 3.2b. The problem will still be statically determinate, since three differential equations of static equilibrium in the plane with proper boundary conditions will be sufficient to compute all three membrane stress resultants without calling upon the midsurface displacements. However, the longitudinal σ_L and the tangential σ_t stresses will no longer be the principal stresses in shell.
   In the design of pressure vessels such an asymmetrically applied load is the wind force on a tall cylindrical column. Using membrane stress theory the computed stresses are identical or close enough to the stresses as computed by elementary beam theory, which is generally used for stress computations in process columns under the wind or earthquake loads. The interested reader will find a concise discussion of asymmetrically loaded membrane shells in refs. 14, 17, 18, and 23.

3.3. CYLINDRICAL SHELLS

Under Uniform Internal Pressure

The cylindrical shell is the most frequently used geometrical shape in pressure vessel design. It is developed by rotating a straight line parallel with the axis of rotation. The meridional radius of curvature $R_L = \infty$ and the second, minimum radius of curvature is the radius of the formed cylinder $R_t = R$. The stresses in a closed-end cylindrical shell under internal pressure P can be computed from the conditions of static equilibrium shown in Fig. 3.4. In the longitudinal direction,

$$2\pi R\sigma_L t = P\pi R^2 \quad \text{or} \quad \sigma_L = PR/2t.$$

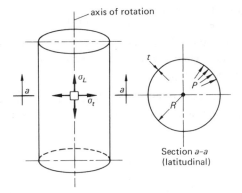

Fig. 3.4.   Stresses in cylindrical (closed end) shell under internal pressure P.

From the first equilibrium equation,

$$(\sigma_L/\infty) + (\sigma_t/R) = P/t,$$

so that

$$\sigma_t = PR/t.$$

The radial growth of the shell is

$$\Delta R = Re_t = (R/E)(\sigma_t - \nu\sigma_L)$$
$$= (PR^2/Et)(1 - \nu/2).$$

There is no end rotation of a cylindrical shell under internal pressure. In the above formulas, E is the modulus of elasticity.

The tangential (governing) stress can be expressed in terms of the inside radius R_i:

$$\sigma_t = PR/t = P(R_i + 0.5t)/t.$$

The shell thickness is therefore

$$t = PR_i/(\sigma_t - 0.5P).$$

The Code stress and shell thickness formulas based on the inside radius approximate the more accurate thick-wall formula of Lamé:

$$SE = (PR_i/t) + 0.6P \quad \text{or} \quad t = PR_i/(SE - 0.6P),$$

using $0.6P$ instead of $0.5P$, where E is the Code weld joint efficiency and S is the allowable Code stress.

Both σ_t and σ_L are principal stresses, without any shear stress on the side of the differential element. However, at other section planes the shear stress appears as in the following example.

Example 3.1. If a spiral-welded pipe (Fig. 3.5) is subjected to internal pressure P, determine the normal and shear forces carried by a linear inch of the butt weld.

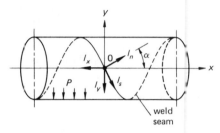

Fig. 3.5.

At point O,

$$l_x = PR/2 \quad \text{and} \quad l_y = PR \text{ lb/in.}$$

Normal tension is

$$l_n = (l_x + l_y)/2 + [(l_x - l_y) \cos 2\alpha]/2$$
$$= (PR/4)(3 - \cos 2\alpha).$$

Shear is

$$l_s = [(l_x - l_y) \sin 2\alpha]/2 = - (PR/4) \sin 2\alpha.$$

As a practical rule the minimum thickness of *carbon steel* cylindrical shells is not less, for fabrication and handling purposes, than the thickness obtained from the following empirical formula:

$$\min. t = [(D_i + 100)/1000] \text{ in.}$$

where D_i is the shell inside diameter in inches.

Under Uniform External Pressure

To compute membrane compressive stresses in a cylindrical shell under uniform external pressure the internal pressure formulas can be used, if the pressure P is

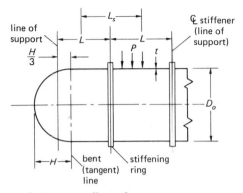

L = distance between two lines of support
L_s = sum of the half of distances from stiffener to the lines of support on each side of stiffener
H = depth of head

Fig. 3.6.

replaced by $-P$. However, thin-wall vessels under external pressure fail at stresses much lower than the yield strength due to instability of the shell. In addition to the physical properties of the construction material at the operating tempera-ture, the principal factors governing the instability and the critical (collapsing) pressure P_c are geometrical: the unsupported shell length L, shell thickness t, and the outisde diameter D_o, assuming that the shell out-of-roundness is within acceptable limits (see Fig. 3.6).

The behavior of thin-wall cylindrical shells under uniform external pressure P differs according to cylinder length.

1. *Very Long Cylinders.* Subjected to a critical pressure P_c, the shell collapses into two lobes by elastic buckling alone (see Fig. 3.7), independent of the supported length L. The stiffeners or the end closures are too far apart to exer-cise any effect on the magnitude of the critical pressure. The only characteristic ratio is t/D_o and the collapsing pressure is given by the following equation [118]:

$$P_c = [2E/(1 - v^2)] (t/D_o)^3$$

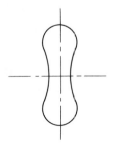

Fig. 3.7 Two-lobe collapse of a pipe under external pressure. The lobes may be irregular.

and for $\nu = 0.3$

$$P_c = 2.2E(t/D_o)^3 . \tag{a}$$

The minimum unsupported length beyond which P_c is independent of L is called the critical length L_c and is expressed by the equation

$$L_c = 1.14(1 - \nu^2)^{1/4} D_o(D_o/t)^{1/2}$$

and for $\nu = 0.3$

$$L_c = 1.11 D_o(D_o/t)^{1/2} .$$

2. *Intermediate Cylinders with* $L < L_c$. If the length L is decreased below the critical distance L_c, the critical pressure P_c and the number of lobes in a complete circumferential belt at collapse n will tend to increase and will become dependent on two characteristic ratios (t/D_o) and (L/D_o).

In the empirical formula for P_c,

$$P_c = KE(t/D_o)^3 \text{ psi}$$

the value of the constant K and the number of lobes n that would produce the minimum P_c for given D_o/t and L/R_o can be read from the charts plotted for carbon steel shells at the room temperature in refs. 105, 120, and 121.

3. *Short Cylinders.* If L becomes short enough the cylinder will fail by plastic yielding alone at high stresses close to the yield strength of the material. The ordinary membrane stress formulas can be used $P_c = (S_y t/R_o)$. This type of failure is common only in heavy-wall cylindrical shells. At this point the influence of L on P_c becomes very small or negligible. In the design of the vacuum thin-wall pressure vessels only the cases 1 and 2 have practical significance.

The theoretical elastic formulas for the critical pressure P_c at which intermediate cylinders would collapse under radial uniform external pressure or under uniform radial combine with uniform axial pressure, are derived in refs. 25 and 118. However, their solutions depend on n, the number of lobes at collapse, and they are cumbersome for a routine design.

To eliminate the dependency on n and to simplify the whole procedure of computation for the wall thickness t for the vacuum vessels at different design temperatures, the Code adopts the following procedure (UG-28). For cylindrical shells with $L < L_c$ the more accurate elastic formulas are replaced by the U.S. Experimental Model Basin formula, which is independent of n and is of sufficient accuracy:

$$P_c = \frac{2.42E}{(1 - \nu^2)^{3/2}} \left[\frac{(t/D_o)^{2.5}}{(L/D_o) - 0.45(t/D_o)^{0.5}} \right]$$

where $0.45(t/D_o)^{0.5}$ can be disregarded and for all practical cases, using $\nu = 0.3$, P_c can be simplified to

$$P_c = 2.80E(t/D_o)^{2.5}/(L/D_o). \tag{b}$$

If equations (a) and (b) are substituted for pressure in the tangential stress formula $P = 2S(t/D_o)$, we obtain a set of two equations giving the tangential stress S_c at collapse:

$$S_c/E = 1.1(t/D_o)^2 \tag{1}$$

for cylinders with $L > L_c$ and

$$S_c/E = 1.40(t/D_o)^{1.5}/(L/D_o) \tag{2}$$

for cylinders with $L \leqslant L_c$. Both equation (1) (vertical) and (2) (slanting) are plotted with S_c/E values as variables on the abscissa and (L/D_o) values as variables on the ordinate for constant ratios (t/D_o) in Fig. UGO-28.0 (see Appendix A2). For better clarity the plot is labeled as (D_o/t).

Since S_c/E is here treated as a variable factor A, this *geometric chart* can be used for all materials. To introduce the particular physical properties of the material, an additional material chart is required, relating the value of the collapsing ratio S_c/E to the collapsing pressure P_c for the particular material. The material chart is actually a strain–stress curve (S_c/E)–$(S_c = P_c D/2t)$ for the material at a design temperature. To obtain the coordinate in terms of the allowable working pressure $P_a = P_c/4$ with a safety factor of four, the hoop formula is again employed:

$$S_c = P_c D_o/2t = (4P_a) D_o/2t$$

and

$$P_a D_o/t = S_c/2.$$

The material chart is then plotted against the same abscissa S_c/E, called factor A (see Fig. UCS-28.2 in Appendix A2), for a specific material and design temperature as ordinate $P_a D_o/t$, called factor B. Factor A ties the two plots together.

Since the modulus of elasticity E starts to reduce for temperatures above $300°F$, the material curves must be drawn for several temperature intervals. To summarize, the geometric chart is used to estimate at which ratios L/D_o and D_o/t the critical stress ratio S_c/E (factor A) will be produced in the shell at some critical pressure.

The *material chart* is used to determine the uniform allowable external pressure P by dividing the factor B by $0.75 D_o/t$, which reduces the safety factor to 3. $P = B/0.75(D_o/t)$ for $D_o/t > 10$.

In practical design when (1) checking the existing vessel for allowable external pressure, compute L/D_o and D_o/t to obtain the factor A from the geometric chart for cylindrical vessels under external pressure. From the material chart determine the factor B and the maximum allowable external pressure is $P = B/0.75(D_o/t)$. When (2) designing a new cylindrical vessel to withstand the external pressure P, assume t (the minimum for carbon steel shells is $t = [(D_i + 100)/1000$ in.]) and compute the factor $B = 0.75P(D_o/t)$. From the material chart find the factor A. Using the factor A move vertically on the geometric chart to the D_o/t line and horizontally to the L/D_o axis. The maximum stiffener spacing or the maximum allowable unsupported length of the cylindrical shell will be $L = (L/D_o) D_o$. Several trials are usually needed to obtain the most economical combination of shell thickness, stiffener size and stiffener spacing.

Stiffener Size. In computing the adequate size of a stiffening ring Levy's formula for buckling of a circular ring under uniform external pressure $(L_s P_c)$ in lb per inch of circumference with the moment of inertia I_s is used:

$$(L_s P_c) = 3 E I_s/R_o^3 = 24 E I_s/D_o^3$$

or

$$I_s = (D_o^2 L_s t/12E)(P_c D_o/2t) = (D_o^2 L_s t/12)(S_c/E).$$

The collapsing strength of the ring is taken as 10 percent higher than the strength of the ordinary shell ($12 \times 1.1 \doteq 14$). The thickness of the shell reinforced by a ring can be increased for computation by the ratio A_s/L_s, where A_s is the area of the stiffener and the effective shell thickness is $(t + A_s/L_s)$:

$$I_s = D_o^2 L_s (t + A_s/L_s) A/14 \text{ in.}^4,$$

where the value of A is read from the material chart for the material used (UG-29). In the above formula, A_s has to be first estimated. The most commonly used shape for ring stiffeners are bars with a rectangular cross section and cut-outs of plate, as shown in Fig. 3.8. Using general formula for the required

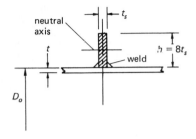

Fig. 3.8.   Stiffener ring for vacuum vessels.

moment of inertia of the stiffener,

$$I = (0.035D_o^4/E)(L_s/D_o)P_c,$$

and substituting $E = 28 \times 10^6$ psi, $P_c = 4P$ and $P = 15$ psi for vacuum vessels,

$$I = (0.075/10^6)D_o^4(L_s/D_o)$$

is the approximate moment of inertia required for the stiffener. The required thickness t_s for a plate stiffener with $h = 8t_s$ (Fig. 3.8) can now be estimated based on the calculated distance L_s:

$$t_s = 0.0065D_o(L_s/D_o)^{1/4} \quad \text{and} \quad A_s = 8t_s^2.$$

If the operating temperature is higher than room temperature, the correction can be made by multiplying t_s by $(28 \times 10^6/E')^{1/4}$, where E' is the value of E at operating temperature. For a preliminary estimate of stiffener sizes of different materials, see ref. 32.

To decrease the size of the stiffener the Code allows shell sections adjacent to the stiffener to be included into the required moment of inertia (Fig. 3.9). Then

$$I_s' = (D_o^2 L_s)(t + A_s/L_s)A/10.9 \text{ in.}^4.$$

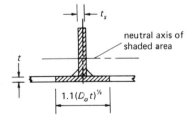

Fig. 3.9.

The stiffener rings are generally placed on the outside of the vessel. On thermally insulated vessels, they must be adequately insulated. If placed inside the vessel, they would obstruct the flow, cleaning etc. If any gaps between stiffener and shell occur, the applicable Code procedure to establish the maximum unsupported length of the gap will apply.

Sometimes the same vessel, reinforced with heavy stiffeners, is subjected occasionally to inside operating pressures. Large discontinuity stresses in the shell can develop adjacent to the stiffeners, and they should be checked [51].

The stiffener-to-shell weld (Fig. 3.8) must carry the load PL_s (lb per inch of the stiffener circumference), and has to be designed accordingly. A fillet weld intermittent on both sides of the stiffener with the Code permitted spacing ($8t$ for external rings) is usually satisfactory.

Only an outline of the analysis of the effects of the external pressure on the cylindrical shells has been presented here—the minimum required to design a vessel under external pressure. The interested reader will find a detailed development and description of the Code charts in ref. 26 and a detailed discussion of cylindrical shells under external pressure in refs. 18, 25, and 118.

3.4. SPHERICAL SHELLS AND HEMISPHERICAL HEADS

Whenever process design or storage conditions permit or high design pressure requires, a spherically shaped vessel is used. Although it is more difficult to fabricate than a cylindrical shell, it requires only half the wall thickness of the cylindrical shell under the same pressure, with minimum exposed surface.

On large-diameter cylindrical vessels, hemispherical heads will introduce negligible discontinuity stresses at junctures. A spherical shell is developed by rotation of a circle around an axis (see Fig. 3.10).

Spherical Shells or Heads Under Internal Pressure.

Both principal radii are the same and the stresses σ_t and σ_L are the same. From the equation specifying the static equilibrium at section a-a in Fig. 3.10,

$$(\sigma_L \cos \alpha)\, t 2\pi R \cos \alpha = \pi R^2 \cos^2 \alpha P$$

we conclude that

$$\sigma_L = PR/2t.$$

But

$$\sigma_L/R + \sigma_t/R = P/t,$$

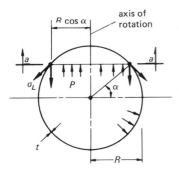

Fig. 3.10.

so that

$$\sigma_t = PR/2t = \sigma_L.$$

The Code stress formula, based on the inside radius and the joint efficiency E:

$$SE = PR_i/2t + 0.1P \quad \text{or} \quad t = PR_i/(2SE - 0.2P),$$

where $0.1P$ is the correction factor. The radial growth ΔR is

$$\Delta R = R(\sigma_t/E - \nu\sigma_L/E) = PR^2/tE[(1 - \nu)/2],$$

where E is the modulus of elasticity.

Both stresses σ_L and σ_t are uniform across the shell thickness in tension. The discontinuity stresses at the cylinder–hemispherical head junction can (per Code Fig. UW-13.1) be minimized by a design taper between the head and the cylindrical shell to such an extent that the stress formula for the sphere thickness can be used to compute the thickness of the hemispherical heads. If the radial growth of the cylinder and connected hemispherical head were the same the discontinuity stress would be eliminated. However, the thickness of the shell would not be fully utilized. If $\Delta R_c = \Delta R_s$ then

$$(PR^2/2Et_c)(2 - \nu) = (PR^2/2Et_s)(1 - \nu)$$

or

$$t_s = t_c(1 - \nu)/(2 - \nu) = 0.41t_c,$$

where

t_s = thickness of the hemispherical head
t_c = thickness of the cylindrical shell.

Spherical Shells and Heads Under External Pressure

The membrane stresses in spherical shells or heads under external pressure can be computed from the stress formulas for the internal pressure P, substituting $-P$ for P. However, as in the case of thin cylindrical shells, thin-wall spherical shells or heads will fail by buckling long before the yield stress in the shell is reached.

To establish the maximum allowable external pressure P_a' in the design of spherical shells, hemispherical heads or sections of a spherical shell large enough to be able to develop characteristic lobes at collapse, the convenient Code procedure has to be used, as described below.

Spherical shells have only one characteristic ratio R_o/t, where R_o is the outside radius and t is the shell thickness. The theoretical collapsing external pressure P_c' for perfectly formed spherical shells is given by

$$P_c' = (t/R_o)^2 \, 2E/[3(1-\nu)]^{1/2}$$

and for $\nu = 0.3$,

$$P_c' = 1.21E(t/R_o)^2.$$

For thin fabricated shells with permissible out of roundness the following equation is used, including the safety factor for the allowable external pressure:

$$P_a' = (E/16)(t/R_o)^2.$$

In order to use the material Code charts for cylindrical shells to determine the allowable external pressure P_a' for spherical shells in the elastic–plastic region, the following adjustment in computations must be made. The material charts were plotted with $S_c/2 = P_a D_o/t = B$ (ordinate) against $S_c/E = A$ a abscissa, where S_c is the critical stress in a cylindrical shell at collapse and P_a is the maximum allowable external pressure for cylindrical shells. The abscissa for a spherical shell can be computed from

$$S_c' = P_c' R_o/2t = 4P_a' R_o/2t = (2R_o/t)[(E/16)(t/R_o)^2]$$

giving

$$S_c'/E = 0.125/(R_o/t) = A.$$

where S_c' is the critical stress in a spherical shell at collapse and P_a' is the maximum allowable external pressure on a spherical shell. However, since $S_c'/E = 2S_c/E$, factor B must be adjusted correspondingly, i.e., multiplied by 2, so that

$2B = P'_a D_o/t$ and the maximum external pressure for spherical shells and heads P'_a is then given by the equation

$$P'_a = B/(R_o/t).$$

Example 3.2. A spherical cover subjected to internal pressure P has a flanged opening, as shown in Fig. 3.11. A uniformly distributed force q (lb/in.) is applied at the top ring flange by a connected pipe. Compute membrane stresses in the cover due to the internal pressure and the force q.

   (a) Membrane stress due to the axisymmetrical load q. The equilibrium equation in the axial direction at angle ϕ is

$$q 2\pi R \sin \phi_0/2\pi R \sin \phi = \sigma_L t \sin \phi$$

or

$$\sigma_L = q \sin \phi_0/t \sin^2 \phi.$$

σ_L is maximum at $\phi = \phi_0$. From $(\sigma_L/R) + (\sigma_t/R) = P/t$ for $P = 0$, obtain

$$\sigma_t = -\sigma_L.$$

The shear and bending stresses at the section at ϕ are disregarded.

   (b) Membrane stresses due to the internal pressure P are

$$\sigma_t = \sigma_L = PR/2t.$$

Both membrane stresses from P and q have to be superposed. The ring flanges on the cover represent a gross structural discontinuity where large secondary

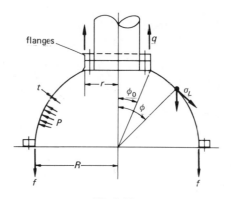

Fig. 3.11.

bending and shear stresses will develop. To evaluate them a more involved analysis would be required. However, away from the ring flanges at a distance greater than $(Rt)^{1/2}$, the simple membrane stresses will be significant.

3.5. SEMIELLIPSOIDAL HEADS

Semiellipsoidal heads are developed by rotation of a semiellipse. Heads with a $2:1$ ratio of major axis R to minor axis h (Fig. 3.12) are the most frequently used end closures in vessel design, particularly for internal pressures above 150 psi and for the bottom heads of tall, slender columns. In the following analysis, semiellipsoidal heads are treated as separate components with no restraints at edges under uniform pressure P.

Under Uniform Internal Pressure

Since both R_L and R_t vary gradually from point to point on the ellipse the stresses σ_L and σ_t vary gradually also. The main radii of curvature R_L and R_t are given by

$$R_t = [R^4/h^2 + (1 - R^2/h^2) x^2]^{1/2}$$

$$R_L = R_t^3 h^2/R^4 .$$

At point 1, $R_t = R_L = R^2/h$ and at point 2, $R_L = h^2/R$ and $R_t = R$. The stresses σ_L and σ_t are the principal stresses, with no shearing stress on the sides of the differential element.

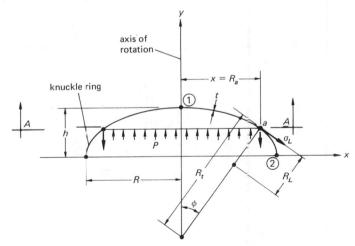

$R_t = R_a/\sin \phi$, tangential radius of curvature.
R_L = longitudinal (meridional) radius of curvature.

Fig. 3.12.   Geometry of a semiellipsoidal head.

The longitudinal stress σ_L can be found from the equilibrium equation written for the latitudinal section A-A in the vertical (axial) direction:

$$2\pi R_a t \sigma_L \sin \phi = \pi R_a^2 P \quad \text{or} \quad \sigma_L = PR_t/2t$$

At point 1,

$$\sigma_L = PR^2/2th$$

and at point 2,

$$\sigma_L = PR/2t.$$

The tangential stress σ_t can be determined from

$$(\sigma_t/R_t) + (\sigma_L/R_L) = P/t \quad \text{or} \quad \sigma_t = (PR_t/t)[1 - (R_t/2R_L)].$$

At point 1,

$$\sigma_t = PR_t/2t$$
$$= PR^2/2th = \sigma_L$$

and at point 2,

$$\sigma_t = (PR/t)[1 - (R^2/2h^2)].$$

From the equation for σ_t at point 2 it can be seen that as long as $R^2/2h^2 < 1$, σ_t remains tensile; if $R^2/2h^2 > 1$ or $R > 1.41h$, σ_t becomes negative, in compression. For standard 2 : 1 ellipsoidal heads with $R = 2h$,

$$\sigma_t = (PR/t)[1 - (4h^2/2h^2)] = -PR/t.$$

The radial displacement at point 2 is

$$\Delta R = R(\sigma_t/E - \nu\sigma_L/E) = (PR^2/2tE)[2 - (R^2/h^2) - \nu].$$

ΔR is positive for $(R/h)^2 + \nu < 2$ and negative for $(R/h)^2 + \nu > 2$. Since there are no discontinuities in a uniformly thick ellipsoidal head the only gross discontinuity is the juncture of the head to the cylindrical shell, which increases the stresses in the knuckle region.

Membrane stress analysis, if used alone in design for various R/h ratios without including the effects of the discontinuity stresses at the head–shell junction, would result in too low a head thickness. To simplify the design procedure the

Code relates the stress design formula for the thickness of ellipsoidal heads to the tangential stress of the cylindrical shell of the radius R, modified by the empirical corrective stress intensification factor K which is based on many tests.

The Code equation for the maximum allowable stress in the head becomes

$$SE = (PD_i/2t) K + 0.1P,$$

where the factor $0.1P$ modifies the stress for use with the inside diameter D_i and factor $K = [2 + (D/2h)^2]/6$ and E is the weld joint efficiency. For $2:1$ ellipsoidal heads $K = 1$ and the head thickness is very nearly equal to that of the connected cylindrical shell. With low discontinuity stresses at the head-shell junction the standard $2:1$ ellipsoidal head is a satisfactory construction at all pressure levels.

In large, thin-wall heads with ratios $R/t > 300$ and $R/h \geqslant 2.5$, a failure in the knuckle region due to the tangential stress σ_t in compression can occur either through elastic buckling (circumferential wrinkles in the meridional direction without any wall thinning) at a stress much less than the yield stress or through plastic buckling (at lower ratios R/h). The main dividing parameter here is the ratio R/t. The division line between the two modes of failure is not clearly defined and has a transition, elastic–plastic range, where both types of failure can occur. The combination of tensile, longitudinal stress σ_L and compressive, tangential stress σ_t becomes significant during hydrotests, when most failures occur.

Unfortunately, there is no completely reliable analysis available for predicting a buckling failure of a semiellipsoidal head under internal pressure at this time. However, the reader will find some guidelines in the refs. 22, 30, 31, 61, and 62.

Under Uniform External Pressure

The membrane stress distribution in an ellipsoidal head due to pressure acting on the convex side can be computed by substituting $-P$ for P in the spherical stress formula for an equivalent radius of the crown section of the head, unless buckling governs.

The design pressure on the convex side of the head is usually much smaller in magnitude than on the concave side, and the Code design procedure (UG33d) has to be used in determining adequate thickness of the head. The procedure is based on the analogy between the maximum allowable compressive stress in the crown region of the head with an equivalent crown radius R and the maximum allowable compressive stress in the externally pressurized spherical shell with the same radius. No buckling will occur in the knuckle region because of the induced high tensile tangential stress before deformation in the crown region occurs.

3.6. TORISPHERICAL HEADS

Torispherical heads have a meridian formed of two circular arcs, a knuckle section with radius r, and a spherical crown segment with crown radius L (see Fig. 3.13).

Under Uniform Internal Pressure

The maximum inside crown radius for Code approved heads equals the outside diameter of the adjacent cylindrical shell. Under internal pressure this would give the same maximum membrane stress in the crown region as in the cylindrical shell. The most commonly used and commercially available torispherical head type is with the minimum knuckle radius equal to 6 percent of L_i. In spite of the similarity between semiellipsoidal and torispherical heads, the sudden change in the radius of curvature (point a in Fig. 3.13) from L to r introduces large discontinuity stresses which are absent in the standard semiellipsoidal heads. However, since they are less expensive to fabricate and the depth of dish H is shorter than in ellipsoidal heads, torispherical heads are quite frequently used for low design pressures (<150 psi). The knuckle region is quite short, and the dis-

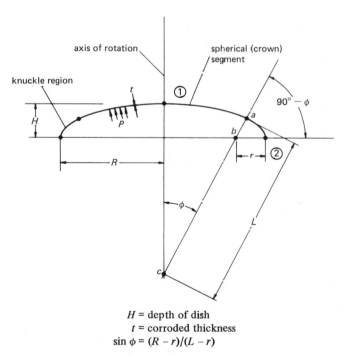

H = depth of dish
t = corroded thickness
$\sin \phi = (R - r)/(L - r)$

Fig. 3.13. Geometry of a torispherical head.

continuity forces at point a have large influence on the discontinuity stresses at point 2, the head–cylinder junction. Also, bending stresses in a knuckle with sharp curvature will be distributed more hyperbolically across the wall thickness than linearly, much as in a curved beam. The local plastic strains induced by high discontinuity stresses at point a tend to cause the knuckle radius to merge more gradually into the crown radius, thus forming a head with a better shape to resist the internal pressure.

As in the design of semiellipsoidal heads, to simplify the procedure of finding an adequate head thickness t the Code introduces an empirical correction factor M into the formula for membrane stress in the crown region, to compensate for the discontinuity stresses at the shell–head junction. The Code formula for the maximum allowable stress in the head becomes

$$SE = PL_i M/2t + 0.1P$$

or

$$t = PL_i M/(2SE - 0.2P),$$

where

$$M = \tfrac{1}{4} \, [3 + (L_i/r_i)^{1/2}]$$

for $r_i = 0.06L_i$, $M = 1.77$ and E is the weld joint efficiency.

Membrane stresses due to inside pressure in the knuckle at point a in Fig. 3.14 are calculated as follows:

$$(\sigma_t/L) + (\sigma_L/r) = P/t$$

$$(\sigma_t/L) + [(PL/2t)/r] = P/t,$$

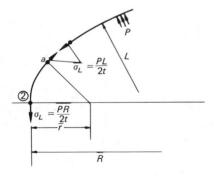

Fig. 3.14. Membrane pressure stresses in the knuckle of a torispherical head.

therefore

$$\sigma_t = (PL/t)[1 - (L/2r)].$$

The above σ_t is the maximum calculated membrane compressive stress in the knuckle at point a, while in the spherical cap the stresses are both in tension and equal to $\sigma_L = \sigma_t = PL/2t$. The actual compressive stress σ_t will be affected by the tensile stress in the adjacent spherical segment, and the final average at point a could be estimated [22] as an average stress equal to

$$\sigma_t = (PL/4t)[3 - (L/r)].$$

Since the tangential compressive stress in the knuckle region of a torispherical head is much larger than that in a semiellipsoidal head, the possibility of failure would seem to be higher. Large, thin-wall torispherical heads are known to collapse by elastic buckling, plastic yielding, or elastic–plastic yield in hydrotests. Because the moduli of elasticity for ordinary and high-strength steels are almost the same, there is no advantage to using high-strength steel for large-diameter, thin-wall torispherical heads.

To predict a possible failure under internal pressure the following approximate formula for the plastic collapse pressure of a torispherical head can be used [29, 107]:

$$P_c/\sigma_y = [0.43 + 7.56(r/d)](t/L) + 34.8[1 - 4.83(r/d)](t/L)^2 - 0.00081$$

where

P_c = collapse pressure, psi
σ_y = yield strength of the used material, psi
r = knuckle radius, in.
d = vessel diameter, in.
t = head thickness, in.
L = crown radius, in.

For standard torispherical heads with $L_i = d_o$ and $r_i = 0.06d_o$, the equation can be written

$$P_c = \sigma_y \{[0.8836 + 24.7149(t/d_o)](t/d_o) - 0.00081\}.$$

The above equation could be used to estimate the collapsing pressure in semi-ellipsoidal heads if values for r and L approximating closely the ellipse are substituted.

Example 3.3. Determine the maximum internal pressure for a standard tori-spherical head, $d_o = 10$ ft, $t = 0.25$ in. minimum, $\sigma_y = 32,000$ psi.

$$d_o/t = 120/0.25 = 480.$$

Code allowable pressure is

$$P = 2SEt/(L_iM + 0.2t)$$
$$= (2 \times 15000 \times 0.25)/(120 \times 1.77 + 2 \times 0.25)$$
$$= 35 \text{ psi}$$

Code yield is

$$P_y = 35 \times 32/15 = 74 \text{ psi}$$

Buckling pressure is

$$P_c = 32,000\{(0.25/120)[0.8836 + 24.7149 \times 0.25/120] - 0.00081\}$$
$$= 36.5 \text{ psi.}$$

If the entire vessel had to be subjected to a hydrotest of $P = 35 \times 1.5 = 53$ psi, an increase in the head thickness would be necessary.

Under Uniform External Pressure

As in the case of semiellipsoidal heads the Code procedure for computing the maximum external allowable pressure P uses the analogy between the compressive stress in the crown region of the head with the allowable compressive stress in the sphere of equivalent radius, and must be followed (UG33e).

3.7. CONICAL HEADS

Under Uniform Internal Pressure

A conical head is generated by the rotation of a straight line intersecting the axis of rotation at an angle α which is the half apex angle of the formed cone. If the conical shell is subjected to a uniform internal pressure P the principal stress σ_t at section a-a in Fig. 3.15 can be determined from the equations

$$\sigma_L/\infty + \sigma_t/R_t = P/t \quad \text{and} \quad \sigma_t = PR_t/t = PR/t \cos \alpha.$$

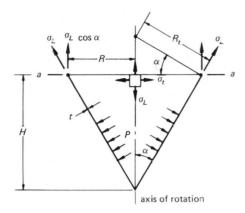

α = half apex angle
$R_t = R/\cos\alpha$, tangential radius of curvature
$R_L = \infty$, longitudinal radius of curvature

Fig. 3.15. Geometry of a conical head.

The equilibrium condition in the vertical direction at section *a-a* yields

$$2\pi R\sigma_L t \cos\alpha = \pi R^2 P \quad \text{and} \quad \sigma_L = PR/2t\cos\alpha.$$

The above stress formulas are the cylindrical formulas where R has been replaced by $R/\cos\alpha$.

The end supporting force $t\sigma_L = PR/2\cos\alpha$ lb/in. at section *a-a* in Fig. 3.15 is shown in the meridional line as required if only membrane stresses are induced in the entire conical head. In actual design, where the conical head is attached to a cylindrical shell, the supporting force $PR/2$ lb/in. is carried by the cylindrical shell, as shown in Fig. 3.16. This arrangement produces an unbalanced force $(PR\tan\alpha)/2$ pointing inward and causing a compressive stress in the region of the junction. Obviously, the larger the angle α, the bigger the inward force. This inward force has to be taken into consideration when discontinuity stresses at

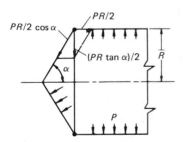

Fig. 3.16. Force diagram at cone–cylinder junction.

cone–cylinder junctions are investigated. The angle α is, therefore, limited in Code design to 30 degrees, and the junction has to be reinforced per Code rules (UA5b and c). Otherwise the Code uses the membrane stress formula for conical shells (UG32g) to determine the maximum stress and thickness of a conical shell with the joint efficiency E:

$$SE = PD_i/2t \cos \alpha + 0.6P,$$

where $0.6P$ is the correction factor accounting for using the inside diameter D_i in the stress formula. If a special analysis is presented the half apex angle α can exceed 30 degrees. However, beyond $\alpha > 60$ degrees the conical shell begins to resemble a shallow shell and finally a circular plate.

The radial growth at section *a-a* is

$$\Delta R = R[(\sigma_t/E) - (\nu\sigma_L/E)] = R^2 P[1 - (\nu/2)]/tE \cos \alpha,$$

where E is the modulus of elasticity.

The rotation of the meridian at section *a-a* is

$$\theta = 3PR \tan \alpha/2Et \cos \alpha.$$

Under External Pressure

The required thickness of the conical head or a conical transition section under external pressure is determined by the same Code procedure as for cylindrical shells, where the actual length of the conical section L in Fig. 3.17 is replaced by

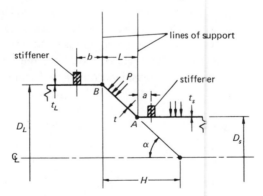

D_L = outside diameter of the large cylinder
D_s = outside diameter of the smaller cylinder

Fig. 3.17.   Conical shell reducer with outside stiffeners.

an equivalent length $L_e = (L/2)[1 + (D_L/D_s)]$ of a cylindrical shell of the diameter D_L $(L_e = H/2$ for conical heads) and the effective wall thickness $t_e = t \cos \alpha$ (UG33f).

Usually the sharp cone–small-diameter cylinder junction, point A in Fig. 3.17, requires reinforcement for discontinuity stresses (UA8c). The stiffening ring for the small cylinder (necessitated by external pressure) should be placed as close to point A as fabrication allows, with the maximum $a = (R_s t_s)^{1/2}/2$. The dimension $(b + L_e)$ should be no larger than the Code allowable distance between the stiffeners for the larger cylinder, with D_L and t_e for the external pressure, unless additional reinforcement at point B is required; where there is additional reinforcement the maximum distance b is equal to $(R_L t_L)^{1/2}/2$. The resultant discontinuity membrane stresses at the point B are in tension and oppose the compressive membrane stresses due to external pressure. The force component $(PR \tan \alpha)/2$ in Fig. 3.17 points outward. The half apex angle α is here limited to 60 degrees. The interested reader will find the development of the Code design method for reducers under external pressure and other loads in ref. 126.

3.8. TOROIDAL SHELLS

A toroid is developed by the rotation of a closed curve, usually a circle about an axis passing outside the generating curve. While an entire toroidal shell, such as an automobile tire, is rarely utilized by itself in the design of pressure vessels, segments of toroidal shells are frequently used as vessel components.

Membrane Stress Analysis of a Toroid with Circular Cross Section under Internal Pressure

Figure 3.18 presents the geometry of a toroid. To find the principal stresses, longitudinal σ_L and tangential σ_t, a ring-shaped toroidal section is isolated and the equilibrium condition between internal pressure P and membrane stress σ_L in the vertical direction is expressed as follows:

$$\pi(R^2 - R_0^2) P = (\sigma_L t \sin \phi) 2\pi R$$

and

$$\sigma_L = P(R^2 - R_0^2)/2tR \sin \phi = (Pr/t)[(R + R_0)/2R] .$$

At point 1, where $R = R_0 - r$,

$$\sigma_L = (Pr/2t)[(2R_0 - r)/(R_0 - r)]$$

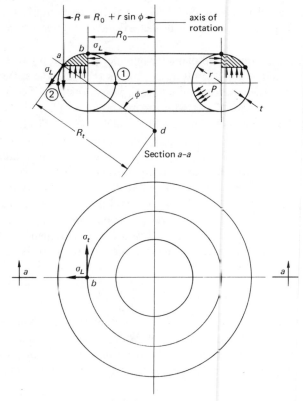

$R_t = R/\sin \phi$, tangential radius of curvature
$R_L = r$, meridional radius of curvature

Fig. 3.18.   Geometry of a toroidal shell.

At point 2, where $R = R_0 + r$,

$$\sigma_L = (Pr/2t)[(2R_0 + r)/(R_0 + r)]$$

At point b, where $R = R_0$,

$$\sigma_L = Pr/t.$$

Note that by the geometry of a toroid σ_L is the longitudinal stress despite the fact that it is tangential to the circle. The meaning of σ_L and σ_t for the toroid is the reverse of that for a straight cylinder.

From the equation

$$[\sigma_t/(R/\sin \phi)] + (\sigma_L/r) = P/t$$

the stress σ_t can be determined:

$$\sigma_t = P(R - R_0)/2t \sin \phi$$

or

$$\sigma_t = Pr/2t.$$

To summarize, both stresses σ_L (variable) and σ_t (constant) are in tension and they are the principal stresses. Stress σ_L at point b is equivalent to the maximum stress in a straight cylinder.

3.9. DESIGN OF CONCENTRIC TORICONICAL REDUCERS UNDER INTERNAL PRESSURE

For a transition section between two coaxial cylindrical shells of different diameters a conical reducer with a knuckle at the larger cylinder and a flare (reintrant knuckle) at the smaller cylinder is very often preferred to a simple conical section without knuckles. The main reason for this is to avoid high discontinuity stresses at the junctures due to the abrupt change in the radius of curvature, particularly at high internal pressures (>150 psig). This can be further aggravated by a poor fit of the cone–cylinder weld joint. The conical transition section with knuckles has both circumferential weld joints away from discontinuities and usually a better alignment with the cylindrical shells; however, it is more expensive to fabricate.

The knuckle at the large cylinder can consist of a ring section cut out of an ellipsoidal, hemispherical, or torispherical head with the same thickness and shape as required for a complete head. More often, both knuckles are fabricated in a form of toroidal rings of the same plate thickness as the conical section.

In the following discussion the required radii for the knuckle r_L and the flare r_s are computed using the principal membrane stresses as governing criteria for the case where the same plate thickness is used for the entire reducer.

The Code specifies only the lower limit for the knuckle radius r_L: "r_{Li} shall not be less than the smaller of $0.12(R_{Li} + t)$ or $3t$ while r_s has no dimensional requirements" (Figs. UG36 and UG32h). The stresses in the conical shell section at point 1 in Fig. 3.19 are given by

$$\sigma_L = PL_1/2t_c = PR_1/2t_c \cos \alpha$$

and from

$$(\sigma_L/\infty) + (\sigma_t/L_1) = P/t_c$$

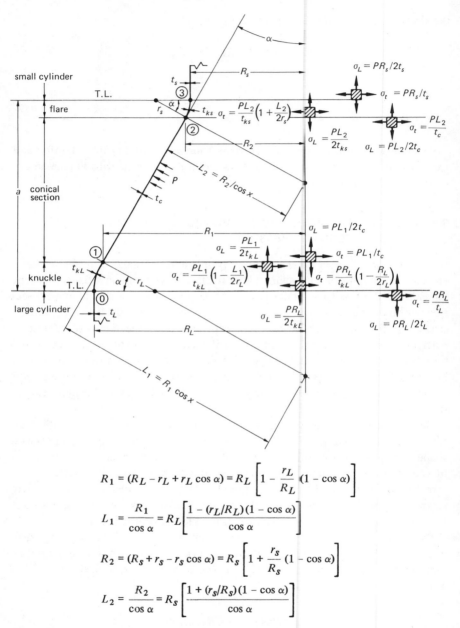

$$R_1 = (R_L - r_L + r_L \cos \alpha) = R_L \left[1 - \frac{r_L}{R_L} (1 - \cos \alpha) \right]$$

$$L_1 = \frac{R_1}{\cos \alpha} = R_L \left[\frac{1 - (r_L/R_L)(1 - \cos \alpha)}{\cos \alpha} \right]$$

$$R_2 = (R_s + r_s - r_s \cos \alpha) = R_s \left[1 + \frac{r_s}{R_s} (1 - \cos \alpha) \right]$$

$$L_2 = \frac{R_2}{\cos \alpha} = R_s \left[\frac{1 + (r_s/R_s)(1 - \cos \alpha)}{\cos \alpha} \right]$$

Fig. 3.19. Membrane stresses in conical transition section with knuckles. All dimensions are to the shell-plate midsurface in corroded condition.

we have

$$\sigma_t = PL_1/t_c = PR_1/t_c \cos \alpha$$

and the cone thickness

$$t_c = PR_1/S_a E \cos \alpha = (PR_L/S_a E)\{[1 - (r_L/R_L) + (r_L \cos \alpha/R_L)]/\cos \alpha\}.$$

Minimum Knuckle Radius r_L

The knuckle at point 1 in Fig. 3.20 will be subjected to the same longitudinal membrane stress σ_L as the conical section on the assumption $t_c = t_{KL}$:

$$\sigma_L = PL_1/2t_{KL}$$

and from

$$(\sigma_L/r_L) + (\sigma_t/L_1) = P/t_{KL}$$

we have

$$\sigma_t = (PL_1/t_{KL})[1 - (L_1/2r_L)].$$

Since L_1 will in practice be always larger than $2r_L$, σ_t will be negative (in compression) membrane stress and maximum at 1. The inward component of σ_L, $\sigma_L \sin \alpha$, will be partially balanced by the internal pressure on the vertical projection of the knuckle ring. The principal radius of curvature L_1 at point 1 is shared by conical and knuckle sections alike and fixed by the geometry of the cone. The second radius of curvature r_L can now be computed in terms of R_L assuming $t_c = t_{KL}$. From stresses σ_t and σ_L by inspection the tangential stress

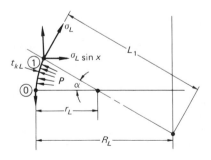

Fig. 3.20.

will govern and can be expressed in terms of R_L at point 1:

$$\sigma_t = (PL_1/t_{KL})[1 - (L_1/2r_L)] = (PR_1/t_{KL} \cos \alpha)[1 - (R_1/2r_L \cos \alpha)]$$
$$= (P/t_{KL} \cos \alpha)(R_L - r_L + r_L \cos \alpha)$$
$$\cdot [(2r_L \cos \alpha - R_L + r_L - r_L \cos \alpha)/2r_L \cos \alpha]$$
$$= -(PR_L/t_{KL})\{[(R_L/r_L) + (r_L \sin^2 \alpha/R_L) - 2]/2 \cos^2 \alpha\}$$

or

$$t_{KL} = (PR_L/-\sigma_t)\{[(R_L/r_L) + (r_L \sin^2 \alpha/R_L) - 2]/2 \cos^2 \alpha\}$$

In replacing $-\sigma_t$ with the maximum allowable compressive stress S_a two criteria must be met: first, the substitution should be in accordance with the intent of the Code rules; and, second, the knuckle radius r_L should be kept to the minimum required, since the larger r_L, the higher the fabrication cost.

The stress $-\sigma_t$ is the maximum calculated compressive membrane stress in the knuckle. However, since there is continuity between the knuckle and the conical section, the final tangential stress in the knuckle at point 1 will be smaller and could be estimated as the average of σ_t in the cone and knuckle [22]. It also has to be remembered that the stresses σ_t and σ_L do not represent the entire stress profile, but are only stress components used as design criteria in absence of a more accurate analysis. The discontinuity stresses would have to be superimposed to obtain the complete state of stress.

Since a knuckle is a section of a torus which in turn approaches a cylinder with an increasing radius R_0 in Fig. 3.18, the maximum allowable compressive stress S_a should be the same as required for cylinders. With small ratios r_L/t_{KL} as used in fabrication, the failure would occur rather in local yielding than elastic buckling. Based on the above reasoning it would seem acceptable to use for the maximum allowable stress $(2S_a E) < S_y$, where S_a is the Code allowable stress in tension and E is the weld joint efficiency. From the condition $t_{KL} = t_c$,

$$(PR_L/2S_a E)\{[(R_L/r_L) + (r_L \sin^2 \alpha/R_L) - 2]/2 \cos^2 \alpha\}$$
$$= (PR_L/S_a E)\{[1 - (r_L/R_L) + (r_L \cos \alpha/R_L)]/\cos \alpha\}.$$

After simplifying,

$$(1 + 4 \cos \alpha - 5 \cos^2 \alpha) r_L^2 - 2R_L(1 + 2 \cos \alpha) r_L + R_L^2 = 0$$

and the minimum radius r_L in terms of the given R_L and the angle α is

$$r_L = R_L/(1 + 5 \cos \alpha) \geqslant 0.12[R_L + (t/2)] \quad \text{or} \quad 3t.$$

For instance, for $\alpha = 30$ degrees, $r_L = 0.188 R_L$.

It can be seen that if the allowable compressive membrane stress $(2S_a E)$ is selected the knuckle radius r_L does not differ a great deal from the minimum r_L as required by the Code. However, if the tangential stress were limited to $1.5 S_a E$, either r_L would become too large or t_{KL} would have to be increased.

Minimum Flare Radius r_s

Similarly, the membrane stresses at point 2 in Fig. 3.21 are

$$\sigma_L = (PL_2/2t_{Ks})$$
$$= (PR_s/2t_{Ks})\{[1 + (r_s/R_s) - (r_s \cos \alpha/R_s)]/\cos \alpha\}$$

and

$$\sigma_t = (PL_2/t_{Ks})[1 + (L_2/2r_s)]$$
$$= [P(R_s + r_s - r_s \cos \alpha)/t_{Ks} \cos \alpha] [(2r_s \cos \alpha + R_s + r_s - r_s \cos \alpha)/2r_s \cos \alpha]$$
$$= (PR_s/t_{Ks})\{[(R_s/r_s) + (r_s \sin^2 \alpha/R_s) + 2]/2 \cos^2 \alpha)\},$$

always in tension.

Using the allowable stress for σ_t equal to $(2S_a E) < S_y$, t_{Ks} is given by

$$t_{Ks} = (PR_s/2S_a E)\{[(R_s/r_s) + (r_s \sin^2 \alpha/R_s) + 2]/2 \cos^2 \alpha\}.$$

From the condition that $t_{Ks} = t_c$ we get

$$(PR_s/2S_a E)\{[(R_s/r_s) + (r_s \sin^2 \alpha/R_s) + 2]/2 \cos^2 \alpha\}$$
$$= (PR_L/S_a E)\{[1 - (r_L/R_L) + (r_L \cos \alpha/R_L)]/\cos \alpha\}$$

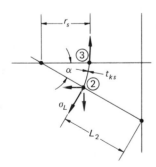

Fig. 3.21.

the minimum flare radius r_s can be computed in terms of R_s and α:

$$\sin^2 \alpha \, r_s^2 + [2R_s - 4 \cos \alpha (R_L - r_L + r_L \cos \alpha)] r_s + R_s^2 = 0,$$

with $(R_L - r_L + r_L \cos \alpha) = R_1 = (S_a E t_c \cos \alpha)/P$, the minimum r_s is given by

$$r_s = \frac{2 S_a E t_c \cos^2 \alpha - P R_s}{P \sin^2 \alpha} - \left[\left(\frac{2 S_a E t_c \cos^2 \alpha - P R_s}{P \sin^2 \alpha} \right)^2 - \frac{R_s^2}{\sin^2 \alpha} \right]^{1/2},$$

where r_s is not made less than $0.12[R_s + (t/2)]$ or $3t$.

Again it has to be remembered that the above analysis is based on membrane pressure stress components only, on which the discontinuity stresses have to be superposed to obtain the entire stress profile.

In order to compute the stresses in cylindrical and conical sections due to external forces (wind, earthquake moments, and weight), standard formulas for bending and compression are used and the resulting stresses are combined with the pressure stresses.

To obtain some criteria about the combined membrane stresses and stresses due to external loads in the knuckle region the following procedure has been successfully used.

First, the longitudinal unit load due to the external loads is determined:

$$l = (\pm 4M/\pi D^2) - (W/\pi D)$$

where M and W are external moment and load at point under investigation, and D is equal to $2R_1$ or to $2R_2$, respectively (see Fig. 3.19). Both values of l should be used in subsequent computations. The longitudinal force in shell due to the external moment M is taken here uniform around circumference.

Second, the "equivalent" internal pressure P_e causing the increased longitudinal stress is computed. From $l = (P'\pi D^2/4)/\pi D$ the additional internal pressure $P' = 4l/D$ is added to the internal pressure P: $P_e = P + 4l/D$.

Third, substituting P_e into the knuckle stress equations the increased membrane stresses σ_t are computed. The stress at point 1 is

$$\sigma_t = (PL_1/t) - (P_e L_1/t)(L_1/2r_L) \leqslant 2S_a E.$$

and the stress at point 2 is

$$\sigma_t = (PL_2/t) + (P_e L_2/t)(L_2/2r_s) \leqslant 2S_a E,$$

where t is the uniform corroded thickness of the conical reducer.

Note 1: The interested reader will find proposed design formulas for the thicknesses of various reducer types in ref. 101.

Note 2: Cone angle. Generally the dimension a in Fig. 3.22 with R_L and R_s given on the process sketch of a process column and the cone half apex angle α, has to be computed. For conveinence the computation is here included.

1. $(R_L - r_L) > (R_s + r_s)$. See Fig. 3.22.

$$\alpha = \gamma + \beta \qquad \tan \beta = \frac{(R_L - r_L) - (R_s + r_s)}{a}$$

$$\sin \gamma = \frac{(r_s + r_L) \cos \beta}{a}$$

2. $(R_L - r_L) < (R_s + r_s)$. Not shown.

$$\alpha = \gamma - \beta \qquad \tan \beta = \frac{(R_s + r_s) - (R_L - r_L)}{a}$$

$$\sin \gamma = \frac{(r_s + r_L) \cos \beta}{a}$$

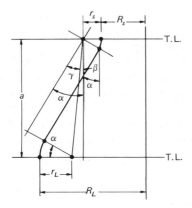

Fig. 3.22.

4
Design of Tall Cylindrical
Self-Supporting Process Columns

4.1. INTRODUCTION

Tall cylindrical process columns built today are self-supporting, i.e., they are supported on cylindrical or conical shells (skirts) with a base ring resting on a concrete foundation and firmly fixed to the foundation by anchor bolts embedded in concrete. Basically, they are designed as cantilever beams.

Several decades ago they were sometimes designed with a guy ring in the top third of the vessel and a number of guy wires at an angle anchored in the ground, a construction still used today for tall stacks. A process column held by guy wires is designed as an overhang beam with one fixed end and one support, with the guy wires exerting an axial force on the column. The thickness of the shell, the size of the foundation, and the number of anchor bolts are reduced by means of the guy wires, as are possible vibration difficulties. However, the additional space required for the guy wires and their anchorage renders this design impractical in modern petrochemical plants.

In some cases the cost of special material required for special tall, slender vessels may be reduced by designing them as guided columns, which are supported at some elevation by an outside service platform built up from the ground and taking part or all of the wind load.

The main dimensions of a process column, i.e., its overall length and inside diameter, as well as the operating pressure and temperature, sizes of nozzle connections, type of trays, and other internals are determined by the process engineer and transferred to the vessel engineer on an analytical data sheet with a sketch of the process column, including all material specifications and corrosion allowance as selected by a metallurgist.

The vessel engineer is responsible for the mechanical design of the process tower. He will originate the computations of the shell thickness, construction of the support, and the size of the anchor bolts, as well as an engineering drawing showing all other required structural details in addition to the above.

The engineering drawing serves the fabricator to prepare all necessary shop working drawings, which are routinely checked by the vessel engineer to secure the correct interpretation of the engineering drawings by the fabricator's technical staff.

4.2. SHELL THICKNESS REQUIRED FOR A COMBINATION OF DESIGN LOADS

The shell thickness of tall slender columns, as computed by the Code formulas, based on the internal or external pressure alone, is usually not sufficient to withstand the combined stresses produced by the operating pressure plus the weight plus the wind or seismic loads. According to Code Division 1, the shell thickness computations are based on the principal stresses—circumferential stress due to the design pressure or longitudinal stress due to the design pressure plus weight plus wind or earthquake load. Since the longitudinal stress increases from top to the bottom of the column, the shell thickness has to be increased below the elevation, where the summation of the longitudinal stresses becomes larger than the circumferential stress due to the design pressure alone.

Detailed stress computations consider the effects of each loading separately. The pressure tangential (circumferential) stress and the combined longitudinal stress are then compared with the Code allowable stress values in tension or compression permitted at the design temperature. In computing the shell thickness at selected elevations it is often advantageous to proceed on the basis of the membrane unit force l (lb/lin. in.) then to use the unit stress σ_L (psi).

Assuming that a tall slender vessel resists the external loads as a cantilever beam the external loads produce bending and shear stresses in the vessel shell. The direct shear having a small value is disregarded in computations. The principal stresses governing the cylindrical vessel thickness required by Code Division 1 (the maximum stress failure theory) are given below.

The tangential stress σ_t due to the pressure is given by

$$\sigma_t = PD/2t \leqslant S_a (\text{psi})$$

where

D = mean corroded shell diameter, in.
P = design pressure, psi
t = corroded thickness of the shell, in.
S_a = Code allowable stress, psi, reduced by joint efficiency if required.

The unit force is $l_t = PD/2 \leqslant S_a t$ (lb/lin. in.).

The combined stress in longitudinal direction σ_L due to the pressure P, dead weight W, and applied moment M, with W and M taken at the elevation under consideration is given as follows.

1. On the windward side,

$$\sigma_L = (PD/4t) + (4M/\pi D^2 t) - (W/\pi Dt) \leqslant S_a \text{ psi}$$
$$l = (PD/4) + (4M/\pi D^2) - (W/\pi D) \leqslant S_a t \text{ lb/lin. in.}$$

and the shell thickness is

$$t = l/S_a = [(PD/4) + (4M/\pi D^2) - (W/\pi D)]/S_a.$$

2. On the leeward side,

$$\sigma_L = (PD/4t) - (4M/\pi D^2 t) - (W/\pi Dt) \leqslant S_a \text{ psi}$$
$$l = (PD/4) - (4M/\pi D^2) - (W/\pi D) \leqslant S_a t \text{ lb/lin. in.}$$

The maximum compressive stress in the shell is induced at the bottom tangent line on the leeward side when the internal pressure is equal to the atmospheric pressure:

$$\sigma_L = -(4M/\pi D^2 t) - (W/\pi Dt) \leqslant S_a$$

or for vacuum vessels:

$$\sigma_L = -(PD/4t) - (4M/\pi D^2 t) - (W/\pi Dt) \leqslant S_a.$$

At this point it will be interesting to compare the Code Division 1 shell thickness, based on the above maximum stress equations and the shell thickness as computed on the basis of maximum shear theory as used in Code Division 2.

The maximum shear stress on the leeward side under internal pressure is

$$\tau = (\sigma_t - \sigma_L)/2 \leqslant S_a/2 \quad \text{or} \quad \sigma_t - \sigma_L \leqslant S_a.$$

Substituting for σ_t and σ_L, we get

$$(PD/2t) - [(PD/4t) - (4M/\pi D^2 t) - (W/\pi Dt)]$$
$$= (PD/4t) + (4M/\pi D^2 t) + (W/\pi Dt) \leqslant S_a,$$

and the shell thickness is therefore

$$t = [(PD/4) + (4M/\pi D^2) + (W/\pi D)]/S_a \text{ in.}$$

which is somewhat larger than the thickness calculated previously by maximum stress theory.

The design procedure for tall process columns of two or more diameters is the same as for a tall slender vessel of a uniform cross section, except that stress analysis of the intersections of the reducers with the cylindrical shell sections should be taken into consideration, as indicated in sections 3.9 and 8.6.

4.3. SUPPORT SKIRTS

Design of the Support Skirt Shell

The support skirts are welded directly to the vessel bottom head or shell, as shown in Fig. 4.1. The factors determining the skirt thickness t_{sk} can be summarized as follows.

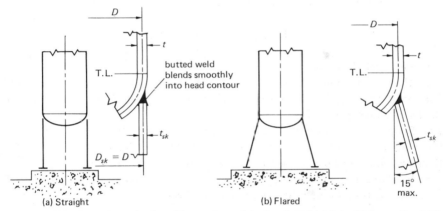

(a) Straight   (b) Flared

Type 1. Skirt butted to the knuckle portion of the head. Weld joint efficiency $E = 0.55$, based on the weld leg equal to t_{sk}.

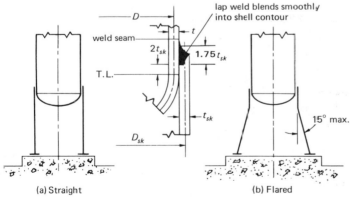

(a) Straight   (b) Flared

Type 2. Skirt lapped to the cylindrical portion of the shell. Weld joint efficiency $E = 0.80$ based on the weld leg equal to t_{sk}.

Fig. 4.1. Types of support skirts and skirt-to-head welds.

1. The maximum longitudinal stress due to the external moment M and weight W at the base is

$$\sigma_L = -(W/\pi D_{sk} t_{sk}) \pm (4M/\pi D_{sk}^2 t_{sk}) \leqslant S_a \text{ psi.}$$

2. The longitudinal compressive stress at the base under test conditions, if the vessel is tested in vertical position, is

$$\sigma_L = -W_T/\pi D_{sk} t_{sk} \leqslant S_a.$$

3. The maximum stress in the skirt-to-head weld, with the weld joint efficiency E depending on the type of the skirt and weld used (see Fig. 4.1), quite often determines the support skirt thickness,

$$t_{sk} = [(W/\pi D_{sk}) + (4M/\pi D_{sk}^2)]/ES_a \text{ in.}$$

where E = weld efficiency.

4. The skirt thickness t_{sk} should be satisfactory for the allowable column deflection; usually t_{sk} for tall towers is chosen not less than the corroded bottom shell section plate thickness.

5. Support skirts for large-diameter vessels, which have to be stress-relieved in the field in a vertical position, must be checked to determine whether the thickness will withstand the weight under high-temperature conditions.

6. If a large access or pipe opening is located in the skirt shell the maximum stress at a section through the opening can be checked:

$$\sigma_L = \pm M/t_{sk} [(\pi D_{sk}^2/4) - (Y D_{sk}/2)] - W/(\pi D_{sk} - Y)t_{sk} \leqslant S_a.$$

If σ_L is too high the opening has to be reinforced (see Fig. 4.2).

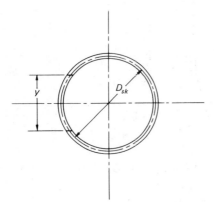

Fig. 4.2. Plan section through a large access opening in the support skirt.

In Fig. 4.1 two typical constructions of support skirts and their attachment welds are shown. Skirt type 1(a) is the most often used design for tall vessels. The centerlines of the cylindrical skirt plate and the corroded shell plate are approximately coincident. If the skirt plate is thicker than the bottom shell plate, the outside diameter of the skirt is made equal to the outside diameter of the bottom shell. If the uplift under the imposed external moment is too high and the anchor bolt spacing becomes too small for the bolt size, the skirt is designed as flared Type 1(b). The localized bending stresses induced in the head by a Type 1(a) skirt are considered acceptable. However, with a Type 1(b) skirt support, the stresses in the head can become excessive and may have to be analyzed more thoroughly.

In Type 2(a) the skirt is attached to the flanged portion of the bottom head in such a way that it does not obstruct an inspection of the head–shell weld seam. This type is more difficult to fabricate and is used mainly for high external loads, high design temperatures, or cyclic operating temperatures. A good fit between the outside diameter of the shell and the inside diameter of the skirt is essential. A flared skirt of Type 2(b) is used for very high columns with extra high external moments.

Skirt Base Design

In Fig. 4.3 two most frequently used skirt base types are illustrated. In Type A the anchor bolts are located off the centerline of the skirt plate. A continuous top stiffening ring is provided to reinforce the support skirt shell against undesirable localized bending stresses.

The disadvantages of the Type B skirt base are the weakening of the skirt shell by the access openings for the bolts and the necessity to check the plate between the openings for buckling. Both dimensions e and c should be kept to a minimum. The bolt circle is equal to the mean diameter of the skirt shell, and the weld connecting the pipe sleeve to the skirt is stressed in shear only because of the holding-down bolt force.

Base Ring Design. To distribute the vertical load over a sufficient area of the concrete foundation a plate base ring is used. In addition it serves also to accommodate the anchor bolts. For the determination of the base ring thickness of both Types A and B the method in the AISC Manual is generally applied. The load is assumed to be uniformly distributed over the entire width b (see Fig. 4.3). The effect of bolt holes and any reinforcement by the vertical stiffeners is disregarded. If the bearing pressure p due to the dead-weight load W combined with the external moment M is

$$p = [(W/\pi D_{sk}) + (4M/\pi D_{sk}^2)]/b \text{ kips/in}^2 .$$

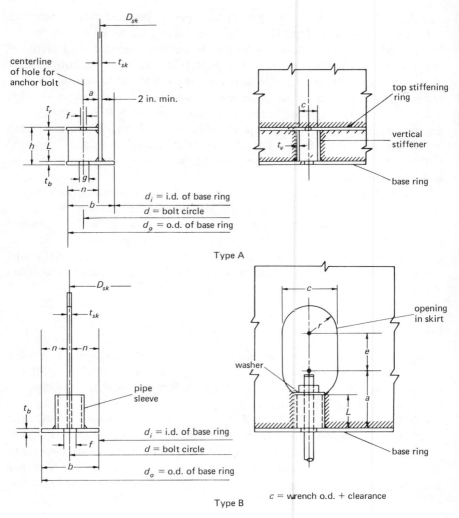

Fig. 4.3.   Types of skirt base.

then the maximum bending stress in the base ring plate is given by

$$\sigma_b = (pn^2/2)/(t_b^2/6) = 3pn^2/t_b^2 \leqslant S_b$$

and the required base plate thickness is

$$t_b = (3pn^2/S_b)^{1/2} \text{ in.,}$$

where n is the dimension in Fig. 4.3 and S_b is the allowable bending stress. For $S_b = 20$ kips/in.^{2}, the formula becomes

$$t_b = (0.15 \, pn^2)^{1/2} \text{ in.}$$

The above formula can be corrected for a lower allowable stress S'_b (psi) as follows:

$$t_b = (0.15 pn^2 \times 20000/S'_b)^{1/2} \text{ in.}$$

The maximum allowable bearing F_b pressure between the base ring and the concrete depends on the compressive strength f'_c of the grade of concrete used. If f'_c is the specified compressive strength (usually between 2 and 5 kips), then, per AISC Manual, Section 1.5.5, the allowable bearing pressure F_b for direct dead load is

$$F_b = 0.25 \, f'_c$$

when the entire area of the concrete support is covered and

$$F_b = 0.375 \, f'_c$$

when only one-third of the area of the concrete base is covered. A value of $F_b = 0.3 \, f'_c$ is generally used and recommended, and the maximum bearing pressure $p \leqslant F_b$. The above value of F_b can be increased by one third for bearing pressures produced by combined dead loads plus wind or earthquake loads (AISC Manual, Section 1.5.6).

Skirt-to-Base Ring Weld. In Type A the holding-down force of the anchor bolts is transferred into the skirt shell by welds connecting the top ring, vertical stiffeners, and the base ring plate, in Type B by the weld between the pipe sleeves and the base ring plate. However, the weld between the skirt and the base ring is assumed to carry the total load and designed as a primary strength weld, preferably continuous.

On the windward side of the vessel the weld has to resist the full uplift load,

$$l_1 = (4M/\pi D_{sk}^2) - (W/\pi D_{sk}) \text{ lb/lin. in.}$$

On the leeward side for the down-load condition any size of weld would theoretically be sufficient. However, since most skirts are so large that their ends cannot be machined accurately enough to produce a uniform bearing, it would seem justified to size the weld to take the full down-load:

$$l_2 = (4M/\pi D_{sk}^2) + (W/\pi D_{sk}) \text{ lb/lin. in.,}$$

assuming that the skirt is not in contact with the base ring, since it would be difficult to estimate the approximate number of contact points. Then w, the size of the weld leg, is

$$w = l_2/f_w \text{ in.,}$$

where the allowable weld unit force f_w (lb/lin. in. of one-inch weld) is

$$f_w = 1.33 \, S_a \times 0.55 \quad \text{for wind or earthquake}$$

$$f_w = 1.20 \, S_a \times 0.55 \quad \text{for the test condition}$$

and S_a is the allowable stress for the skirt base plate or skirt shell plate, whichever is smaller.

Top Stiffening Ring. The continuous top ring with anchor bolt holes and welded to the skirt shell as shown in Fig. 4.3, Type A, helps to distribute the bolt holding down reactions more evenly into the skirt shell. To determine the required thickness t_r would require an involved stress analysis, including the base ring, vertical stiffeners, and skirt section. The thickness t_r could be computed from the concentrated load at the edges of the bolt holes. A rectangular plate with dimensions n and c can be *roughly* approximated by a beam with the longer ends fixed, load P on the plate, and minimum section modulus $Z = t_r^2 (n - f)/6$. Based on the above simplifying assumptions the thickness t_r can be estimated by

$$t_r = [Pc/4S_b(n - f)]^{1/2} \text{ in.,}$$

where P is the maximum bolt load (approximated by 1.25 times bolt stress area A_b times the bolt allowable stress S_a) and S_b is the allowable stress in bending for the top ring material.

Vertical Stiffeners. The vertical stiffeners in Fig. 4.3, Type A, are welded to the skirt and top and bottom base rings. The distance c between stiffeners is kept to the minimum that fabrication allows for the computed bolt size. On the leeward side the stiffeners are stressed in compression and their thickness can be computed as that of a plate supported on three sides with one side free. A simpler and more conservative approach is to treat the stiffener as a plate column. If the thickness t_V is assumed, then the maximum allowable unit stress for axially loaded columns per the AISC column formula should not exceed

$$P/2a \leqslant 17000 - 0.485 \ (L/r)^2 \ \text{psi},$$

where

$a = t_V(n - 0.25)$, the cross-sectional area of one stiffener, in.
L = length of the stiffener, in.
$r = (I/a)^{1/2} = 0.289 t_V$, the radius of gyration of the stiffener, in.
P = the maximum bolt load, lb.
$(n - 0.25)$ = effective width of the stiffener, in.

The thickness t_V is usually between $\frac{1}{2}$ and $1\frac{1}{4}$ in., depending on bolt size. Where no uplift or only a very small uplift results from external loads the top ring section can be omitted between bolts and the design reduces to a bolting chair. The minimum size for bolts ($\frac{3}{4}$ to 1 in.) is selected and the anchor bolts serve merely to locate the vessel in place.

Skirt Material

The most frequently used materials for support skirts are carbon steels, A283 gr. C for thicknesses up to $\frac{5}{8}$ in. and A285 gr. C for thicknesses $\frac{5}{8}$ in. and above. Since this is a very important structural part the allowable stresses used are the same as for the pressure parts, without increasing them by one-third for combined stresses due to weight and wind.

The heavy base rings are also fabricated from A285 gr. C with yield strength S_y = 30000 psi. Since the allowable bending stress for structural steels with S_y = 36000 psi in the AISC base plate formula is 20000 psi, the allowable bending stress for A285C to be used in the AISC formula can be determined as follows:

$$S = 20000 \times 30/36 = 16700 \ \text{psi} \quad \text{weight only}$$

$$S = 1.33 \times 16700 = 22200 \ \text{psi} \quad \text{weight plus wind}$$

$$S = 1.2 \times 16700 = 20000 \ \text{psi} \quad \text{test weight}$$

Higher allowable stresses are acceptable for the base ring than for the skirt shell because minor deformations of the base ring from overstressing would not cause any damage.

4.4. ANCHOR BOLTS

Self-supporting columns must be safely fixed to the supporting concrete foundations with adequately sized anchor bolts embedded in the concrete to prevent overturning or excessive swaying from lateral wind or earthquake loads. To compute the tension stress in the bolts and their required size and number one of three methods can be applied: (1) a simplified method, using generalized design conditions and ignoring dynamic effects and necessary preloading of bolts; (2) a more complete method, considering initial preload on bolts; and (3) disregarding initial preload on bolts.

Simplified Method

The forces and the moments acting on a tall slender vessel are shown in Fig. 4.4 as assumed for the anchor bolt analysis.

Assuming that the column will rotate about the axis y in Fig. 4.4, the maximum uplift force F per bolt due to the outside moment M is determined as follows. The maximum tension on the bolt circumference in lb. per lin. in. is given by

$$T = (M/Z_L) - (W/C) = (4M/\pi d^2) - (W/\pi d) \text{ lb/in.}$$

If $4M/\pi d^2$ is larger than $W/\pi d$ there is a positive uplift force inducing tension stress, with magnitude depending on the distance x spanned by half the anchor

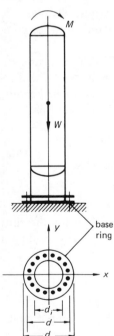

A_b = bolt tensile stress area, in.^{2}
M = overturning moment at base due to wind or earthquake, lb-in.
W = weight of the vessel, operating (W_o) or erection (W_e), lb
d = bolt circle diameter, in.
d_o = outside diameter of the base ring, in.
d_i = inside diameter of the base ring, in.
N = total number of anchor bolts in multiples of 4
$Z_L = \pi d^2/4$, linear section modulus of the bolt circle, in.^{2}
$C = \pi d$, circumference of the bolt circle, in.
x = distance of an anchor bolt from the neutral axis y, in.
S_a = allowable design stress for the anchor bolts, psi

Fig. 4.4.

bolts. The maximum force F on the bolt at distance $x = d/2$ from the axis y is

$$F = T\pi d/N = (4M/dN) - (W/N) \text{ lb/bolt}$$

and the required bolt area is

$$A_b = [(4M/d) - W]/NS_a.$$

The anchor bolts are not designed for the horizontal shear force since it is clearly counteracted by the friction between the base ring and the foundation.

Obviously, this approach is simplified, since it does not try to establish more accurately the actual design conditions, the initial bolt preload, and the dynamic effect of the external overturning moment. However, it is generally accepted for sizing the anchor bolts and, if a relatively conservative allowable stress S_a is used in the design, it gives acceptable results and is very simple to apply.

Initial Preload in Bolts Considered

Since wind and earthquake loads are essentially dynamic, initial tightening of the bolt nuts is required to reduce the variable stress range or any other impact effect on the nut under operating conditions.

The maximum force per bolt resulting from the combined action of the over-turning moment M and the initial preload could be found by an approximate analysis with the following design conditions assumed, as follows.

1. The initial bolt preload together with the weight of the vessel is large enough to maintain a compressive pressure between the vessel base plate and the concrete pedestal under design loads.

2. As long as this compression exists at the contact area the skirt base and the pedestal behave as a continuous structure and the support base will rotate about the neutral axis of the contact area (axis y in Fig. 4.4).

Under moment M the maximum and the minimum pressure on the contact area becomes

$$S_c = (NF_i/A_c) + (W/A_c) \pm (Md_o/2I_c),$$

where

$$A_c = \pi(d_o^2 - d_i^2)/4$$
$$I_c = \pi(d_o^4 - d_i^4)/64$$

and F_i is the initial bolt load due to pretightening of the bolt nut, lb/bolt. The minimum F_i to maintain the compression is when $S_c = 0$:

$$F_i = (Md_o A_c/2NI_c) - (W/N) = 8d_o M/N(d_o^2 + d_i^2) - (W/N)$$

or, after simplification,

$$F_i \doteq (4M/Nd) - (W/N).$$

The external moment M in the above equation is the maximum design moment. Actually, it fluctuates from zero to some maximum value. The maximum force on the bolt due to this moment is given by

$$F_a = (4M/Nd) - (W/N).$$

However, since the anchor bolt forms an elastic joint with the skirt base on the concrete pedestal, pretightened by the force F_i, only a part of F_a will be carried by the bolt [33].

The total variable load F per bolt under moment M can be visualized from the diagram in Fig. 4.5. Using the diagram, the following equation can be derived:

$$c'F_c = c'F_i - e'(F - F_i)$$

$$F_c = F_i - (F - F_i)e'/c'$$

$$F = F_a + F_c = F_a + F_i - (F - F_i)e'/c'$$

$$= F_i + [c'/(c' + e')]F_a = F_i + CF_a$$

and the minimum tensile stress in the bolt is

$$S = (F_i/A_b) + (CF_a/A_b).$$

Substituting the components for F_a we get

$$F = F_i + F_{am} \pm KF_{ar},$$

where K is the stress concentration factor from Table 4.1 applied to the variable component only.

The factor $C = c'/(c' + e')$ for hard elastic joints is quite small. The actual value of C would be difficult to evaluate. However, if for instance c' is one sixth of

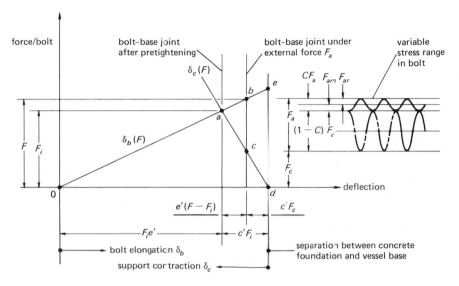

F_i = initial bolt preload, lb/bolt
F_a = applied operating load, lb/bolt
F = combined total load on a bolt, lb/bolt
F_c = compressive load on the vessel support under a bolt; at point a, $F_c = F_i$
$e' = l_b/A_b E_b$, rate of elongation of the bolt, in./lb
c' = rate of compression of the combined supports, in./lb. If the concrete foundation is assumed to be rigid, $c' = l_b/A_s E_s$, where l_b is the bolt length, E_b is the modulus of elasticity of the bolt material, A_s is the support steel area in compression, and E_s is the modulus of elasticity of the vessel base
$F_{am} = CF_a/2$, steady load component of CF_a in bolt
$F_{ar} = \pm CF_a/2$, variable load component of CF_a in bolt

Fig. 4.5. Force–deflection diagram for anchor bolt and support base.

Table 4.1. Stress Concentration Factor K for
Threaded (American Standard) Steel Fasteners
Subject to Tensile Loads.

	K	
	ROLLED FASTENER	CUT FASTENER
Annealed	2.2	2.8
Quenched and tempered	3.0	3.8

e', then $C = 0.143$ and the maximum tensile stress at the thread root would be

$$S = F/A_b$$
$$= [F_i + (0.143\, F_a/2) + (2.2 \times 0.143\, F_a/2)]/A_b$$
$$= (F_i/A_b) + (0.229\, F_a/A_b).$$

From Fig. 4.5 it is obvious that as long as the force F_a does not exceed the initial load F_i there is not a substantial increase in the total bolt load F. It is also apparent how important the initial preload is in reducing of the variable force F_{ar} and thus inproving the fatigue life of the bolt. Any force load per bolt due to the excessive moment M beyond the separation point d is algebraically additive to the total force F (de).

The required bolt area A_b is given by

$$A_b = F/S_a.$$

Neglecting the term with the factor C, the expression for the required bolt area reduces to

$$A_b = F_i/S_a$$

or approximately

$$A_b = \left[\frac{(4M/d) - (W)}{NS_a}\right],$$

which is the form as computed by the simplified method.

From the above discussion it would seem that the simplified method combined with conservative design bolt stresses (because some terms were disregarded) represents a satisfactory design.

An additional design safety factor is the use of the highest wind velocity that occurs at infrequent intervals. If a loss of initial tension in the bolts occurs because of creep under heavy winds, the bolts have to be retightened, all bolts equally as possible.

The minimum approximate *initial torque for the required F_i* is given by

$$T = F_i(1 + N'd_b)/2\pi N' \text{ lb.-in.}$$

where

N' = number of threads per inch of the bolt
d_b = nominal bolt diameter, in.

The proof load of a bolt is usually defined as the maximum load a bolt can withstand without a permanent deformation. The preloading of bolts up to 75 percent of proof load is quite common; however, there is a possibility that too high a preload can impair the fatigue strength of the bolts. Yielding of the bolt can be caused if too high a mean stress $(F_i + F_{am})$ combines with a high alternating stress (KF_{ar}). If an anchor bolt fails, it is usually by elongation.

Example 4.1. A tall column (Fig. 4.6) is subjected to a wind moment at base $M = 1.2 \times 10^6$ ft-lb. The operating weight W_o is equal to 200,000 lb and the erection weight W_e is 150,000 lb. Determine the required size and number of anchor bolts. As a first estimate, assume $N = 12$.

Required bolt area under operating conditions is

$$A_b = [(4M/d) - W]/NS_a$$

$$= (1/12 \times 15,000) \, [(4 \times 1.2 \times 10^6 \times 12/79) - 200,000]$$

$$= 2.93 \text{ in.}^2$$

Use 12 $2\frac{1}{4}$-in.-diameter bolts with stress area $A_b = 3.25$ in.^{2} Maximum bolt stress under erection dead load is

$$S = (729,114 - 150,000)/(12 \times 3.25) = 14,850 \text{ psi.}$$

Bolt spacing is

$$(\pi \times 79)/12 = 20.7 \text{ in.}$$

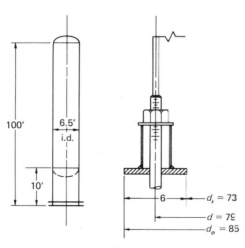

Fig. 4.6.

Maximum bearing pressure on contact area is

$$S_c = (NF_i/A_c) + (W/A_c) + (Md_o/2I_c)$$
$$= (12 \times 3.25 \times 15{,}000)/1490 + (200{,}000/1490)$$
$$+ (1.2 \times 10^6 \times 12 \times 42.5)/(1.168 \times 10^6)$$
$$= 393 + 134 + 525 = 1052 \text{ psi.}$$

Maximum allowable bearing pressure on concrete is, using $f'_c = 3000$ psi.

$$F_b = 0.3 \times 3000 = 900 \text{ psi} \qquad \text{weight only}$$
$$F_b = 1.33 \times 900 = 1200 \text{ psi} \qquad \text{weight plus wind.}$$

Initial Tension in Bolts Neglected

Nuts are assumed to be hand tight on bolts, with no initial load on bolts. Because of the column weight a certain compression between the base of the column and the concrete foundation will exist, but can be partially overcome by the external moment M. The vessel base will partially separate from the pedestal and will be held down by some anchor bolts on the windward side.

The moment M is resisted by a portion of the anchor bolts and the bearing pressure between the vessel base and the foundation. The total bolt area can be replaced by an equivalent area of steel cylindrical shell:

$$A_s = NA_b = t_s \pi d \qquad \text{or} \qquad t_s = NA_b/\pi d$$

as shown in Fig. 4.7.

The result is a composite cantilever beam with an unknown area in tension and unknown concrete area in compression (section a-a of Fig. 4.7), subjected to given external loads W and M.

The design procedure used to compute the required steel area A_s is similar to that employed in reinforced beam theory.

First, the location of the neutral axis is established. The strains in the steel shell $e_s = S_s/E_s$ and in concrete $e_c = S_c/E_c$ are directly proportional to the distances from the neutral axis. If S_s and S_c are the maximum allowable stresses, then from similar triangles (Fig. 4.7, section a-a) we get

$$e_s/e_c = (S_s/E_s)/(S_c/E_c) = (1 - k)d/kd$$

and

$$k = 1/[1 + (S_s/nS_c)].$$

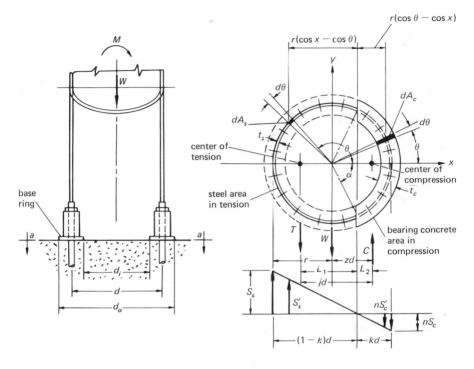

Section *a–a*

E_s = modulus of elasticity of steel
E_c = modulus of elasticity of concrete
$n = E_s/E_c$
S_s = maximum allowable tensile stress for steel
S_c = maximum allowable compressive stress for concrete
S_s' = stress in steel at variable angle θ from α to π, in tension
S_c' = stress in concrete at variable angle θ from 0 to α, in compression
r = bolt circle radius

Fig. 4.7.

This equation will specify k in terms of maximum allowable stresses S_s and S_c under the assumption that both maximum allowable stresses occur simultaneously at the most remote elements from the neutral axis. Further,

$$\cos \alpha = [(d/2) - kd]/(d/2) = 1 - 2k$$

and

$$S_s' = S_s (\cos \alpha - \cos \theta)/(1 + \cos \alpha)$$
$$S_c' = S_c (\cos \theta - \cos \alpha)/(1 - \cos \alpha).$$

Since the allowable stress S_s is larger than nS_c, the position of the neutral axis will move from the y axis toward the compression area. This will produce a larger resisting steel area and a lower total number of anchor bolts than computed by previous methods.

The summation of elemental forces on the steel area in tension, $2\int_\alpha^\pi S_s' \, dA$, can be represented by a force resultant T located at the center of tension at a distance L_1 from the neutral axis. After integration,

$$T = S_s t_s d [(\pi - \alpha) \cos \alpha + \sin \alpha]/(1 + \cos \alpha).$$

From the summation of the moments of the elemental forces about the neutral axis the distance L_1 can be determined:

$$L_1 = \frac{2}{T} \int_\alpha^\pi S_s' \, dA_s \, r \, (\cos \alpha - \cos \theta)$$

$$= \frac{d}{2} \left[\frac{(\pi - \alpha) \cos^2 \alpha + (3 \sin \alpha \cos \alpha)/2 + (\pi - \alpha)/2}{(\pi - \alpha) \cos \alpha + \sin \alpha} \right].$$

Similarly the bearing force resultant C and the distance L_2 can be computed:

$$C = t_c dS_c [(\sin \alpha - \alpha \cos \alpha)/(1 - \cos \alpha)]$$

and

$$L_2 = (d/2) [\alpha \cos^2 \alpha - (3 \sin \alpha \cos \alpha)/2 + (\alpha/2)]/(\sin \alpha - \alpha \cos \alpha).$$

The distance jd between the forces T and C is given by

$$jd = L_1 + L_2.$$

The distance zd of the force C from the y axis is given by

$$zd = L_2 + r \cos \alpha.$$

From the summation of moments of all known forces about the location of the force C, the value of the force T can now be computed in terms of the known loads acting on the vessel:

$$M - W(zd) - T(jd) = 0$$

$$T = [M - W(zd)]/(jd).$$

Substituting the above result into the previous expression for T, the thickness of the steel shell can be computed:

$$t_s = \frac{1}{S_s d} \left[\frac{M - W(zd)}{(jd)} \right] \left[\frac{1 + \cos \alpha}{(\pi - \alpha) \cos \alpha + \sin \alpha} \right]$$

and *the required bolt area* is

$$A_b = t_s \pi d / N \text{ in.}^2 .$$

From the summation of the vertical forces acting on the vessel, $T + W - C = 0$, the minimum base ring width t_c can be determined:

$$t_c = [(T + W)/S_s d] [(1 - \cos \alpha)/(\sin \alpha - \alpha \cos \alpha)] .$$

To accommodate the anchor bolts t_c is usually made larger than required by the above equation for the computed location of the neutral axis. To maintain the position of the neutral axis and the validity of computations the area A_s is increased in direct proportion to A_c. However, the stresses in bolts will decrease in inverse proportion to A_c. If t_c and A_c only are increased the neutral axis shifts toward the compression side and A_s becomes oversized and can be decreased. In order to find the minimum A_s a redesign would be required.

Only an outline of this method has been presented here. The interested reader will find any additional information in refs. 42–45. The main shortcoming of the method is the omission of the initial preload in the bolts and the treatment of the loads as static. If high allowable stresses are used, the design can lead to unconservative results, and this is probably why this approach has not been generally used.

In practice, a considerable preload always has to be applied, and the bolts used should be of sufficient size to permit retightening in excess of minimum design requirements in order to prevent large stress fluctuations. On a few occasions in the writer's experience, excessive swaying of process towers developed simply because the anchor bolts were not pulled up tight enough by the erection crews.

Allowable Stresses for Carbon Steel Anchor Bolts

The standard material used for anchor bolts is carbon steel type A307 gr. A or B, with a minimum tensile strength of 60,000 psi. Bolt threads, standard coarse thread series or eight-thread series, should preferably be rolled, with forged nuts of heavy series. Sometimes a corrosion allowance of $\frac{1}{16} - \frac{1}{8}$ in. is specified and two nuts (one as a lock nut) per bolt are used where frequent heavy winds can be expected.

The allowable stress in tension for the threaded part of the anchor bolt is a matter of safe design, subject to engineering judgment. The allowable stress for A307 bolts in the AISC Manual is 20,000 psi. This high allowable stress would require an accurate stress analysis with well defined static loads. Noting that

wind and earthquake loads are dynamic loads,
there is always a possibility of overload,
a failure of a process column would cause a large loss of property,
bolt material is comparatively cheap,
inspection of the anchor bolts is often inadequate, and
any repair in the field is very costly,

it is reasonable to accept lower allowable stresses.

It is a common engineering practice to select the allowable stresses for carbon steel anchor bolts as follows:

$$S_a = 15,000 \text{ psi} \qquad \text{under operating conditions}$$

$$S_a = 15,000 \times 1.2$$

$$= 18,000 \text{ psi} \qquad \text{under empty (erection) condition.}$$

Types of Anchor Bolt

Some typical details of installed anchor bolts are shown in Figs. 4.8 and 4.9. The required length L of the bolt embedded in the concrete foundation is designed by the civil engineer. It has a holding power equal to the full tensile strength of the bolt. Minimum bolt spacing is usually set at 10 times nominal bold diameter. The dimension h is usually specified as 12 in.; the longer the free length of the bolt, the higher its resilience, i.e., its ability to absorb an impact load without permanent deformation.

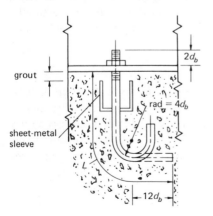

Fig. 4.8. Anchor bolt for small vessel.

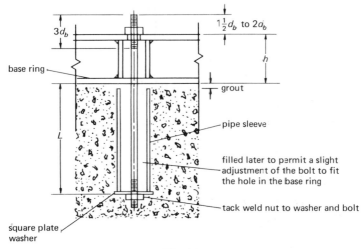

$3d_b$

base ring

$1\frac{1}{2}d_b$ to $2d_b$

h

grout

pipe sleeve

L

filled later to permit a slight
adjustment of the bolt to fit
the hole in the base ring

tack weld nut to washer and bolt

square plate
washer

Fig. 4.9.

4.5. WIND-INDUCED DEFLECTIONS OF TALL COLUMNS

A sustained wind pressure will cause a tall column to deflect with the wind. The magnitude of the deflection may seriously influence the performance of a process column and has to be limited to a certain value. If too small a deflection at the top of the column is specified by the client's specifications, the shell thickness must be increased and the price of the tower will increase. Most engineering specifications ask for a maximum deflection of 6 in. per 100 ft of column height.

The deflection at the top of tall slender columns ($H/D \geqslant 15$) is routinely checked. Deflection computations are often computerized; however, a need may arise for such calculations to be made by hand.

No matter what short-cut analytical method is used, the computations are comparatively lengthy and subject to error. The method should be flexible enough to permit easy inclusion of such variables as shell thickness, modulus of elasticity due to changes in operating temperature, and wind pressure above the ground. The method should also be as simple as possible and permit the use of the results from previous computations of the wind moments and the wind loads for the determination of the shell thicknesses.

There are a number of methods which can be used.

The method of superposition can be applied here to advantage. The vessel is assumed to be a cantilever beam firmly fixed to the concrete pedestal. The effect of foundation movement is considered negligible. The six basic formulas used in this method are shown in Table. 4.2.

Consider the cantilever beam of Fig. 4.10. Using the formulas in Table 4.2 the individual deflections in Fig. 4.10 can be easily evaluated:

$$y_1 = \frac{L_1^2}{EI_1} \left[\frac{W_1 L_1}{8} + \frac{Q_1 L_1}{3} + \frac{M_1}{2} \right]$$

$$y_2 = \frac{L_2^2}{EI_2} \left[\frac{W_2 L_2}{8} + \frac{Q_2 L_2}{3} + \frac{M_2}{2} \right]$$

$$\Delta_{1-2} = \frac{L_1 L_2}{EI_2} \left[\frac{W_2 L_2}{6} + \frac{Q_2 L_2}{2} + M_2 \right]$$

$$y_3 = \frac{L_3^2}{EI_3} \left[\frac{W_3 L_3}{8} + \frac{Q_3 L_3}{3} + \frac{M_3}{2} \right]$$

$$\Delta_{1-3} = \frac{(L_1 + L_2) L_3}{EI_3} \left[\frac{W_3 L_3}{6} + \frac{Q_3 L_3}{2} + M_3 \right]$$

$$y_4 = \frac{L_4^2}{EI_4} \left[\frac{W_4 L_4}{8} + \frac{Q_4 L_4}{3} + \frac{M_4}{2} \right]$$

$$\Delta_{1-4} = \frac{(L_1 + L_2 + L_3) L_4}{EI_4} \left[\frac{W_4 L_4}{6} + \frac{Q_4 L_4}{2} + M_4 \right]$$

$$\vdots$$

$$y_i = \frac{L_i^2}{EI_i} \left[\frac{W_i L_i}{8} + \frac{Q_i L_i}{3} + \frac{M_i}{2} \right]$$

$$\Delta_{1-i} = \frac{(L_1 + L_2 + \cdots + L_{i-1}) L_i}{EI_i} \left[\frac{W_i L_i}{6} + \frac{Q_i L_i}{2} + M_i \right]$$

Table 4.2. Cantilever Beam Formulas.

TYPE OF LOAD	END ANGLE θ AT FREE END	END DEFLECTION y
$W = wL$	$\dfrac{WL^2}{6EI}$	$\dfrac{WL^3}{8EI}$
Q	$\dfrac{QL^2}{2EI}$	$\dfrac{QL^3}{3EI}$
M	$\dfrac{ML}{EI}$	$\dfrac{ML^2}{2EI}$

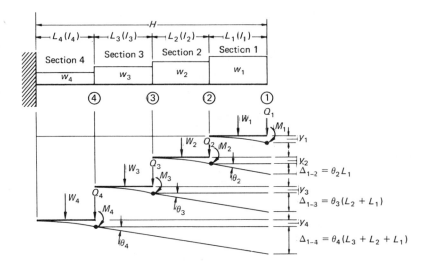

W_i = total wind load on sections i. $W_1 = w_1 L_1$, $W_2 = w_2 L_2$, and so on, where w_1, w_2, are the unit wind loads in lb per ft of column height.

Q_i = shear load at point i, equal to the summation of all W-loads above the point. Q_1 = piping thrust, if any. $Q_2 = W_1 + Q_1, Q_3 = W_2 + Q_2$, etc.

M_i = end moment acting at points i, due to the loads above the point. M_1 = piping moment, if any. $M_2 = (W_1 L_1/2) + Q_1 L_1 + M_1$; $M_3 = (W_2 L_2/2) + Q_2 L_2 + M_2$; etc.

y_i = deflection of section i, at point i due to the wind loads W_i, end load Q_i, and end moment M_i.

$\Delta_{1-2}, \Delta_{1-3}$, etc. = deflections at point 1 due to the end angles θ_2, θ_3, etc. induced by the wind loads W_2, W_3, etc., end loads Q_2, Q_3, etc., and end moments M_2, M_3, etc., at the ends of the respective sections.

Fig. 4.10. Schematic diagram of load and deflection for sections of a cantilever beam using the superposition method.

The total deflection at the top of the column is

$$y = y_1 + y_2 + y_3 + \cdots + y_i + \Delta_{1-2} + \Delta_{1-3} + \Delta_{1-4} + \cdots + \Delta_{1-i}.$$

If the deflection y is based on modulus of elasticity E at a design temperature, the deflection y' at another operating temperature (with E') is equal to y multiplied by E/E'.

Example 4.2. Evaluate the deflection at the top of the process column with the wind loads shown in Fig. 4.11. Modulus of elasticity is $E = 30 \times 10^6 \times 144$ psf.

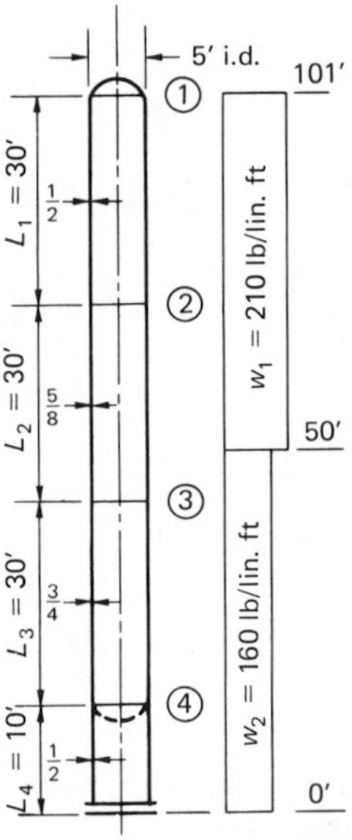

	W_i (lb)	Q_i (lb)	M_i (lb-ft)	$I = \pi R_i^3 t_i$ (ft^4)	EI_i (lb-ft²)	Section (i)
①	6,500			2.1	9.05×10^9	1
		6,500	97,500			
②	5,800			2.64	11.4×10^9	2
		12,300	379,500			
③	4,800			3.18	13.76×10^9	3
		17,100	821,500			
④	1,600	18,700	1,293,500	2.1	9.05×10^9	4

Fig. 4.11. Design data for Example 4.2.

Deflection at the top of the column in feet is figured as follows.

$$y_1 = \frac{6,500 \times 30^3}{8 \times 2.1 \times 30 \times 10^6 \times 144} = 0.0024 \text{ ft}$$

$$y_2 = \frac{30^2}{11.4 \times 10^9}\left[\frac{5,800 \times 30}{8} + \frac{6,500 \times 30}{3} + \frac{97,500}{2}\right] = 0.0091$$

$$\Delta_{1-2} = \frac{30 \times 30}{11.4 \times 10^9}\left[\frac{5,800 \times 30}{6} + \frac{6,500 \times 30}{2} + 97,500\right] = 0.0177$$

$$y_3 = \frac{30^2}{13.76 \times 10^9}\left[\frac{4,800 \times 30}{8} + \frac{12,300 \times 30}{3} + \frac{379,500}{2}\right] = 0.0216$$

$$\Delta_{1-3} = \frac{60 \times 30}{13.76 \times 10^9}\left[\frac{4,800 \times 30}{6} + \frac{12,300 \times 30}{2} + 379,500\right] = 0.0770$$

$$y_4 = \frac{10^2}{9.05 \times 10^9}\left[\frac{1,600 \times 10}{8} + \frac{17,100 \times 10}{3} + \frac{821,500}{2}\right] = 0.0052$$

$$\Delta_{1-4} = \frac{90 \times 10}{9.05 \times 10^9}\left[\frac{1,600 \times 10}{6} + \frac{17,100 \times 10}{2} + 821,500\right] = 0.0905$$

Total deflection $y = 0.2235$ ft
$= 2.7$ in.

4.6. WIND-INDUCED VIBRATIONS

Introduction

Early in this century it was observed that some tall cylindrical chimneys were vibrating with a high frequency perpendicular to the direction of comparatively low-velocity winds. Since this type of vibration proved to be destructive to such structures the phenomenon attracted the attention of number of investigators who attempted to formulate a rational theory applicable to the economical and safe design of tall, slender cylindrical structures.

This type of vibration affects tall, unlined steel stacks far more frequently than it does process columns. There are several reasons for the relative immunity of tall, slender cylindrical vessels. For instance, they have relatively thick walls, with high first frequency and resisting strength. Also, external piping, ladders, and platforms tend to disrupt wind flow around the vessel, and vibrations tend to be damped by the operating liquid in the vessel and in connected piping. However, some process towers can be subject to wind-induced vibrations, and may have to be checked for them.

Tall stacks are routinely checked for cross-wind vibration. A detailed examination is usually required, however, to determine whether a tall process column needs to be checked for vibration-inducing aerodynamic forces.

Basic Aerodynamic Principles

In the evaluation of forces acting on a long, stationary, circular cylinder normal to a uniform stream of wind, a detailed study of the wind flow around the cylinder is necessary. Fig. 4.12 shows the fluid pattern around a stationary cylinder if the fluid is ideal, frictionless, and incompressible. The flow pattern is symmetrical from front to back of the cylinder. The same flow is observed with real fluids at very low Reynolds numbers.

The Reynolds number Re is a dimensionless parameter expressing the ratio of the inertia forces to the viscous forces in a particular fluid flow. When Re is small, the inertia forces of the stream are small, the stream behaves in a more

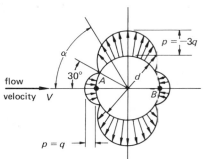

Fig. 4.12. Pressure distribution around a long cylinder for ideal fluids. A and B are front and rear stagnation points, respectively.

viscous manner and the flow tends to be laminar. When Re is large, the inertia forces are large compared to viscous forces and the fluid flow becomes turbulent.

If Re has the same numerical value for two geometrically similar systems (including the position of the object in the stream) the fluid flow pattern in both systems can be expected to be similar.

For atmospheric conditions of air flow around a long smooth cylinder the Reynolds number Re is given by equation

$$Re = \rho V d / \mu$$

where

$\rho = 2.4 \times 10^{-3}$, slugs/ft^3 or lb sec^2/ft^4, air density
V = wind velocity, ft/sec
d = outside diameter of the cylinder, ft
$\mu = 3.8 \times 10^{-7}$, lb sec/ft^2 or slug/sec ft, viscosity of air.

When the above numerical values are substituted into the above equation for Re, we get

$$Re = 6{,}320Vd, \quad \text{for wind velocity } V \text{ in ft/sec, or}$$

$$Re = 9{,}270Vd, \quad \text{for wind velocity } V \text{ in mph.}$$

Development of the Force on a Cylinder Parallel to the Direction of Flow

The pressure distribution around the cylinder in Fig. 4.12 is given by $p = q(1 - 4\sin^2 \alpha)$, where $q = \rho V^2/2$ is the dynamic pressure at the stagnation point A where the flow velocity is zero. The net force resultant acting on a unit length of the cylinder is zero.

In actual air flow the friction and the air compressibility will change the flow pattern around the cylinder and the pressure distribution. The streamlines do not follow the body contour back to the rear. The pressure pattern will resemble more that shown in Fig. 4.13, indicating the development of a force acting on

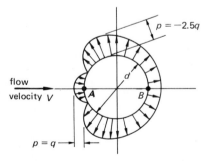

Fig. 4.13. Schematic illustration of pressure around a long cylinder for actual flows at higher Re values.

the cylinder parallel to the direction of the flow, called *drag*. The pressure on the front section remains nearly the same as in Fig. 4.12. However in the rear, wake action changes positive pressure to negative. The pressures in the rear and in the front are no longer balanced. The unit pressure acting on the projected area normal to the direction of the flow is equal:

$$D = C_D \rho V^2 / 2 \text{ (psf)},$$

where C_D is the drag coefficient, experimentally determined at various Re values. The value of C_D is well established and, except for very low Re values, can be taken as equal to 0.9–1.1, until the critical range $Re > 10^5$, beyond which it drops to 0.3–0.5.

The total drag force D consists of the friction component D_f, predominant at lower Re values, and the pressure component D_p due to the pressure resultant on the cylinder. Usually, D_p for blunt bodies will be much larger than the force D_f. The drag force is dealt with in the design of pressure vessels as the wind load.

Development of the Force on a Cylinder Transverse to the Direction of Flow

Because of friction a boundary layer of retarded fluid forms on the cylinder surface. A separation of the flow occurs near the rear of the cylinder and a wake, a turbulent region with a lower flow velocity than the surrounding free stream, develops. With high Re, the separation points of the flow move forward toward the transverse diameter of the cylinder and the wake width increases (Fig. 4.14).

By friction the higher-velocity enclosing layers set the lower velocity fluid immediately behind the separation points in rotation. Two symmetrical vortices with opposite rotational velocities develop, and as the wind velocity increases, the friction force between the free stream and vortices increases.

At some point one vortex breaks away and passes downstream, disintegrating in the free stream [Fig. 4.14(d)].

Up to the critical value of $Re < 5 \times 10^5$ the vortices will form and shed alternately on either side of the cylinder, giving rise while forming to a transverse force L (psf) on the cylinder, at a frequency of vortex shedding f_V. Beyond this critical Re value the separation points move backward, the wake contracts and becomes entirely turbulent, and the vortex periodicity vanishes.

Von Karman was the first to analyze the formation of such vortices mathematically, and found that maximum stability in the flow of vortices occurred when the vortices were shed alternately from the sides of a cylinder, as shown in Fig. 4.15. The vortices moving downstream at a speed somewhat less than that of the free stream form a so-called Von Karman vortex street or trail.

The cross thrust force L, like the drag force, can be expressed conveniently as

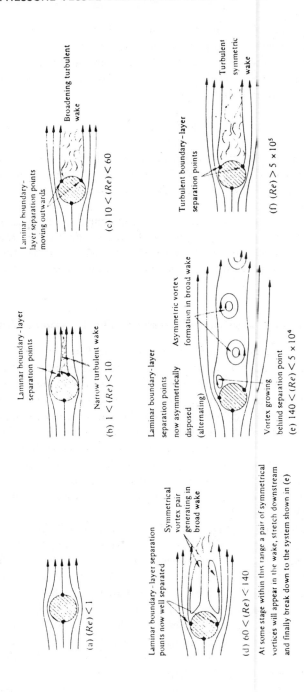

Fig. 4.14. Flow of air past a stationary cylinder perpendicular to the direction of the flow. The *Re* numbers are only approximate. (Reprinted from ref. 123, courtesy of Edward Arnold Publishers Ltd., London.)

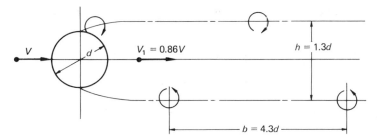

Fig. 4.15.   Vortex trail in the wake of a cylinder.

a part of the stagnation pressure $q = \rho V^2/2$:

$$L = C_L \rho V^2/2 \quad \text{(psf of the projected area)}.$$

Unfortunately, in case of a cylinder no firm value of C_L has been established. Values for C_L ranging from 0.2 to as high 1.7 are given in technical sources. Calculated C_L values based on field observations are usually low, while the value of $C_L = 1.7$ is based on a mathematical analysis. Values of C_L between 0.2 to 1.0 appear to be generally accepted and used in practice; a value of 1.0 is considered conservative. For practical engineering computations, values of $C_L = 0.4$–0.6 are recommended here.

Critical Wind Velocity

The first original investigation of the vortex forming and shedding phenomenon was done in 1878 by Strouhal, who found the following relationship:

$$S = f_V d/V$$

where

    S = a dimensionless parameter, called the Strouhal number
    f_V = shedding frequency of vortices, cps
    d = diameter of the cylinder perpendicular to the flow of the wind, ft
    V = wind velocity, fps.

The value of S varies with the values of Re. However, in the range of Re values between 10^3 and 10^5 the value of S remains nearly stable, equal to between 0.18 and 0.21, and is generally taken as equal to 0.20 in computations. For values of $Re > 10^5$ the Strouhal number increases rapidly to 0.35.

The first critical wind velocity V_1 is the wind velocity at which the shedding frequency of wind vortices f_V is equal to the first natural frequency of the column f. It can be computed from the Strouhal number, substituting $f = 1/T$ for f_V:

$$V_1 = fd/0.2 \text{ (fps)}$$
$$V_1 = 3.40d/T \text{ mph.}$$

The second critical wind velocity can be taken as $V_2 = 6.25V_1$.

Basic Vibrational Principles; Magnification Factor

When the vibration of stacks was first observed it was ascribed to resonance. According to this theory, if the frequency f_V of the cross thrust force L coincides with one of the natural frequencies f of the cylinder, however it may be supported, a forced resonant vibration results.

The stresses induced by such oscillations can build up and increase over those calculated from static wind load analysis.

At this point some consideration of the basic vibrational principles applicable to this problem is appropriate. A tall vessel, a cantilever beam, represents a system with a distributed mass of infinitely many vibrational freedoms and natural periods of vibration, each of them accompanied by a distinct mode or vibrational curve. However, the prevailing period of vibration will be the longest, called fundamental or first, which is of main interest. The general conclusions pertaining to the resonance case can be derived by using a system with a single degree of freedom, as shown in Fig. 4.16.

The differential equation for damped, forced harmonic motion of a system with a single degree of freedom expresses the displacement x of mass m between its equilibrium position and its instantaneous location as a function of time t:

$$m\ddot{x} + c\dot{x} + kx = F \cos \omega t$$

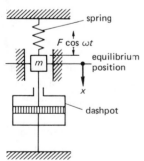

Fig. 4.16. System with a single degree of freedom subject to viscous damping and an externally imposed harmonic force.

where

$m = W/g$ is the mass

$c =$ damping factor, resistance to motion in pounds when the velocity is equal to one inch per second, lb-sec/in.

$k =$ scale of a weightless spring, force in pounds required to deflect the spring by one inch, lb/in.

$F \cos \omega t =$ harmonic periodic force impressed on the system, lb. It is maximum when $t = 0$. It could also be taken as $F \sin \omega t$, with the minimum when $t = 0$, depending on the desired approximation

$\omega =$ the circular frequency of the impressed force F, rad./sec.

The general solution of this differential equation consists of two parts superposed on each other.

1. *The complementary function* is the general solution of the homogeneous equation (the right-hand side set equal to zero). With two independent constants, the complementary function represents the free damped vibration of the system and is given by:

$$x = e^{-(c/2m)\,t}(C_1 \sin \omega_d t + C_2 \cos \omega_d t)$$

where $\omega_d = [(k/m) - (c/2m)^2]^{1/2}$, is the damped natural circular frequency and C_1 and C_2 are constants which depend on two initial conditions. For small damping coefficients c, as in the case of process columns, ω_d can be replaced by $\omega_n = (k/m)^{1/2}$, the natural circular frequency for a vibrating system with no damping and no impressed force and described by the equation $m\ddot{x} + kx = 0$.

Free vibration disappears because of damping. Very light damping will allow many oscillations to occur, but they will gradually lessen in amplitude and become negligible. However, it is quite possible that the intensity of the induced free vibration, for instance during an earthquake, is such that it will cause damage to the vessel before it has the time to die out.

2. *The particular integral* is the simplest possible solution of the equation, with no integration constant, specifying the forced oscillation of the system:

$$X = \frac{F/k}{\left\{\left[1 - \left(\dfrac{\omega}{\omega_n}\right)^2\right]^2 + \left[2\,\dfrac{c}{c_c}\,\dfrac{\omega}{\omega_n}\right]^2\right\}^{1/2}} \cos (\omega t - \phi),$$

where

$c_c = 2\,(mk)^{1/2}$ is the critical damping factor, representing the dividing line between overdamping $(c/c_c > 1$, nonvibrating motion) and underdamping ($c/c_c < 1$, harmonic vibration). The damping factor c of a tall, slender pressure vessel is usually only few percent of c_c

$F/k = X_s$ is the deflection of the system due to the maximum impressed force F when acting as a static force

ϕ = constant phase angle.

The maximum actual amplitude x of forced vibration can be obtained by multiplying the statical deflection X_s by the fraction

$$\text{M.F.} = 1/\{[1 - (\omega/\omega_n)^2]^2 + [(2c/c_c)(\omega/\omega_n)]^2\}^{1/2}$$

called the magnification factor. For small values of c/c_c ($\leqslant 0.2$) the maximum value of M.F. will be reached very nearly at $\omega = \omega_n$:

$$\text{Maximum M.F.} = c_c/2c.$$

In practical computations the damping coefficients c and c_c of the system are not known, not readily measurable, and difficult to estimate with sufficient accuracy. For free vibration with damping less than critical, the ratio of two consecutive maximum amplitudes is (see Fig. 4.17)

$$\frac{x_{n+1}}{x_n} = \frac{\exp\{-(c/2m)[t + (2\pi/\omega_d)]\}}{\exp[-(c/2m)t]}$$

$$= \exp(-\pi c/m\omega_d) = e^{-\delta}$$

where δ is known as the logarithmic decrement and is given by $\delta = \pi c/m\omega_d$. Expressed in terms of c/c_c, and substituting for ω_d and $mk = c_c^2/4$,

$$\delta = (\pi c/m)[k/m - (c/2m)^2]^{1/2} = 2\pi(c/c_c)/[1 - (c/c_c)]^{1/2}$$

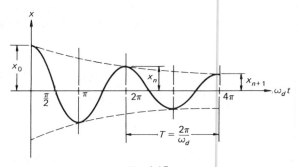

Fig. 4.17.

and for small values of c/c_c,

$$\delta = 2\pi(c/c_c).$$

The magnification factor at resonance can now be expressed as follows:

$$\text{Maximum M.F.} = \pi/\delta.$$

Some investigators consider this value too high to apply against cylindrical structures, and suggest that a maximum M.F. equal to $1/\delta$ is more in line with their actual measurements.

Logarithmic Decrement δ

Damping in a system can also be defined as the rate at which the material absorbs energy under a cyclic load. The energy is dissipated through heat by internal damping due to microscopic plastic action in the material. Damping of a vibrating column would include, in addition to internal damping, friction losses between the shell and air, any energy losses due to resistance of the external piping and internals, and conditions at the base support.

Since $x_{n+1}/x_n = e^{-\delta} \doteq 1 - \delta$, the logarithmic decrement δ can be expressed by a simple ratio,

$$\delta \doteq (x_{n+1} - x_n)/x_n$$

which can be interpreted as the percentage decay per cycle in the amplitude of a freely vibrating vessel and thus provides a means of measuring the vibration decay rate of a column in field, since the value of two consecutive amplitudes x_n and x_{n+1} can be determined by field measurements.

However, δ lacks full significance, since it describes conditions away from the point of resonance and is susceptible to errors in measurement. Most available data, however, are presented in terms of δ, although its reliable values for typical process towers and stacks are not as extensive as it would be desirable.

Values of δ for welded towers and stacks and magnification factors based on field measurements are given in Table 4.3. Unless otherwise specified, the values of δ and M.F. in columns II and III can be used in most applications.

Since a deflection is directly proportional to the applied force, the maximum transverse force per unit area of the projected surface of a cylinder at resonance can be expressed by

$$L = (C_L \times \text{M.F.})\, \rho V_1^2/2 \text{ (psf)}.$$

For the first estimate a value of $(C_L \times \text{M.F.}) = (0.5 \times 60)$ can be used.

Table 4.3. Average Values for δ and M.F.

	I		II		III	
	SOFT SOILS		STIFF SOILS		ROCK, VERY STIFF	
	δ	M.F.	δ	M.F.	δ	M.F.
Tall process columns	0.126	25	0.080	40	0.052	60
Unlined stacks	0.105	30	0.052	60	0.035	90
Lined stacks	0.314	10	0.105	30	0.070	45

Source: Ref. 48.

Conflict between Forced and Self-Excited Vibration Theories

All values essential to the aerodynamic design analysis of towers and stacks, such as the coefficient C_L, the Strouhal number S, and particularly the damping data, are subjects of wide differences of opinion among various investigators. Furthermore, the question has been raised whether forced vibration theory should be used as the basic design-governing theory in the aerodynamic design of stacks and towers.

According to forced vibration theory the alternating transverse force acting on a cylinder in an air flow originates from vortex shedding, which is independent of the motion of the cylinder. This force exists at all wind velocities at which the von Karman vortices are formed. Since the frequency of forced vibrations is equal to the frequency of the impressed force, this would mean, that the stack would vibrate to some degree at a frequency proportional to wind velocity. At the critical wind velocity the frequency of vortex shedding f_V equals to the natural frequency of the stack or tower f and resonance occurs. The principal reason this theory has not been generally accepted is that the frequencies of stack vibrations as observed in the field have nearly always been found to be approximately constant at the value of the natural frequencies of the stacks. This is in poor agreement with forced vibration theory.

To explain this discrepancy other theories have been advanced, particularly self-excitation theory. A self-excited vibration should take place only near the natural frequency of the structure. In self-excited vibration the impressed alternating force sustaining the motion is created by the motion itself and, when the motion of the object stops, the impressed force disappears. Once excited to vibration the cylinder controls the frequency of the impressed force [36, 47]. Vibration would continue through a wide range of wind velocities, and a particular critical wind velocity would not exist. If the amount of energy extracted by the cylinder from the air exceeds damping, the amplitude of oscillations will increase until, in practice, an equilibrium between damping and the energy input develops. However, a full satisfactory theory for a stack oscillating with self-excited vibrations has not been mathematically worked out by its proponents.

It would seem that at low-Re flows the cylinder is dynamically stable, subject only to forced vibrations; at the higher-Re flows a possibility of self-excited vibration exists. From a large number of field observations it can be concluded that the first peak amplitudes appear at the critical wind velocity V_1, corresponding to a Strouhal number of approximately 0.2 pointing out the forced vibration theory as the basic excitation. The formation of secondary peak amplitudes at higher wind velocities in some cases may lead to modification of forced vibration theory, particularly after the behavior of streams around elastically deflecting cylinders at higher Re values is better known.

For a vessel designer the most significant fact is that the peak vibrations as predicted by forced vibration theory are not inconsistent with the observed data and can form a basis for the mathematical checking of a stack or tower for wind-induced vibrations.

Design Method Based on Forced Vibration

The first critical wind velocity V_1 is given by

$$V_1 = 3.40d/T \text{ mph}$$

where d is the outside diameter of the vessel in feet, and T is the first period of vibration in second per cycle.

The second critical velocity can be taken as $V_2 = 6.25V_1$.

The maximum unit pressure L transverse to the wind direction at resonance on the projected cylinder area is

$$L = (C_L \times \text{M.F.}) \rho V_1^2/2 \text{ psf.}$$

Assuming that wind pressure and tower mass effects are concentrated in the vessel top section ($\frac{1}{2}$ to $\frac{1}{4}$ of H depending on exposure) and column stiffness effects are limited to the vessel bottom section, the action of the total pressure acting on a cylindrical tall slender vessel at resonance can then be approximated by its resultant equivalent static force at the top of the column:

$$F = (C_L \times \text{M.F.})(\rho V_1^2/2)(d \times H/3).$$

When V_1 is in mph,

$$F = 0.00086(C_L \times \text{M.F.})(d \times H \times V_1^2).$$

The equivalent force F can be used in computing the maximum nominal stresses at various elevations of the shell and also the maximum resonance amplitude at the top of the vessel. The induced stresses must be superposed on the stresses due to weight and operating pressure. The total combined stress should not exceed the maximum allowable stress for the shell material. If the combined stresses are too high and a possibility of fatigue failure exists, changes in the design have to be made.

No vibrational analysis would be completed without an attempt to evaluate the effects of fatigue. The stress induced in the bottom shell section, welds, and skirt support is completely reversed in bending with the stress range $S_R = 2S$. The following relationship between the number of cycles to failure N and the stress range S_R is used as a fatigue curve [74]:

$$N = (K/\beta S_R)^n,$$

where n and K are material constants and β is the stress intensification factor. For carbon steel with an ultimate strength of 60,000 psi, $n = 5$ and $K = 780,000$. The value of β will vary according the type of construction, weld and inspection. Some suggested values for β are:

$\beta = 1.2$ for a shell plate with a smooth finish
$\beta = 1.8$ for butt-weld joints
$\beta = 3$ for fillet weld or incomplete penetration groove weld with the root un-
    sealed between the bottom head and the support skirt.

The fatigue life expectency Le in hours of vibration in a steady wind at the resonant velocity will be given by

$$Le = NT/3,600.$$

The safe service life of continuous vibration is

$$Le/\text{S.F.}$$

No precise criteria for the safety factor S.F. can be given. However, the minimum safe service life should be at least equal to the sum of probable time periods of steady wind of resonant velocity at the job site. A safety factor of 10–15, as applied to cycles to failure, could be accepted for this type of load and structure.

From the above discussion it is quite obvious that, with so many assumed variable factors, the results of design computations can only be interpreted as approximations, the starting point for the evaluation of the vessel behavior.

Corrective Measures against Vibration

The most commonly used preventive measure against vibration is a spiral vortex spoiler welded around the top third of the stack. Another corrective measure for an existing structure subject to excessive vibration would be installation of permanent guys. Special external damping devices are expensive, impractical, and in practice not very effective. They do not seem to represent a satisfactory solution.

As in any mechanical design the cheapest and most effective vibration-preventive measures can be accomplished by a careful analysis during the design stage, even at a higher initial cost.

Ovaling

Since mostly unlined stacks are affected by this resonance phenomenon only a brief discussion will be offered here. The cross aerodynamic forces can induce oscillating deformations of the upper section of stacks called *ovaling* or *breathing*. The pressure pulsations which cause ovaling of a stack occur once per vortex formation. If the natural frequency of a cylindrical shell taken as a circular ring coincides with the vortex shedding frequency, the shell will have tendency to flatten periodically into an ellipse, with the direction of the major axis varying from perpendicular to parallel to the wind direction.

The lowest natural period of a cylinder taken as a ring is given by

$$T = 2\pi/\omega_n = (2\pi/2.68)(mR^4/EI)^{1/2}$$

where

m = the mass per unit length of the ring; for steel, m = 1 in. $\times$ 1 in. $\times$ t $\times$ 0.28/386)

E = modulus of elasticity, 27×10^6 psi

$I = t^3/12$, moment of inertia, in.^{4}.

Substituting the above values, we get

$$T = d^2/660t$$

where d is the mean cylinder diameter in feet and thickness t in inches.

The resonant wind velocity which theoretically would induce ovaling is

$$V_1 = 3.40d/2T \quad \text{or} \quad V_1 = 1,120t/d \text{ mph.}$$

Reinforcing rings with a required section modulus are added in the top third of a stack to secure it from ovaling. A spiral vortex spoiler can be used in place of the rings [124].

Limiting Values for a Vibrational Analysis

It would be advantageous to be able to determine in advance from a single, simple parameter whether a vibrational analysis is necessary. Unfortunately, no such parameter is available. Most investigations have been done on tall stacks, which can be straight cylindrical, tapered, or half-tapered; lined with gunite or unlined; and of welded or riveted construction. The weight is nearly evenly distributed along the height, and there is no connected process piping involved.

Empirical parameters intended for stacks, therefore cannot blindly be applied to tall towers. As already mentioned, process columns are not afflicted with cross-wind vibrations very often. However, given the trend to higher, slender vessels, the possibility of vibration will likely be of increasing importance. To obtain more reliable criteria, more research laboratory work and experimental work on full-scale, field-erected, high vessels are needed.

The following can be used as general guidelines in deciding whether a vibrational analysis is required.

1. The upper limit of the critical wind velocities V_1 and V_2 can be limited to 60 mph, since a very small possibility of a sustained wind velocity beyond this limit will exist. This value also is as high as any known wind velocity at which cross vibrations have been recorded.

2. If V_1 is greater than the design wind velocity used in static pressure computations, no further check is required. Only if V_1 falls in the range of prevailing wind velocities at the site area is a further investigation justified.

3. The limiting minimum height-to-diameter ratios H/d for vibrational analysis are:

unlined stacks:	$H/d \geqslant 13$
lined stacks	$H/d \geqslant 15$
process columns:	$H/d \geqslant 15$

4. Possible criteria, relating to total weight W (lb), height H (ft), and average diameter of the top half of the vessel d (ft) are suggested in ref. 49 to establish the need for a vibrational analysis, as follows:

$$W/Hd^2 \leqslant 20, \quad \text{analysis must be performed}$$
$$20 < W/Hd^2 \leqslant 25, \quad \text{analysis should be performed}$$
$$25 < W/Hd^2, \quad \text{analysis need not be performed.}$$

5. According to an often used rule of thumb, complete vibratory analysis of a stack or a tower is not required if the total force on the stack or tower caused by the wind of critical velocity V_1 (V_2 if it is less than the wind design velocity) does not exceed $\frac{1}{15}$ of the operating (corroded) weight W_o or

$$\tfrac{1}{2} \rho V^2 Hd/W_o \leqslant 0.067.$$

In the above formula, H is the total height of a uniform-diameter stack or tower. For a flared stack, H can be replaced by an equivalent height

$$H_e = H \text{ (cylinder)} + H/2 \text{ (cone)}.$$

For lined stacks, W_o includes the weight of the internal lining.

4.7. FIRST NATURAL PERIOD OF VIBRATION

The first natural period of vibration T of tall slender vessels is an important criterion in design for wind or earthquake loads. For a tall, slender cylindrical tower of uniform diameter and thickness, equivalent to a fixed-end cantilever beam, T is given (in seconds per one complete cycle) by

$$T = (1/0.560) \, (wH^4/gEI)^{1/2}$$

where

$g = 32.2 \text{ ft/sec}^2$, gravitational acceleration
E = modulus of elasticity, 29×10^6 psi
H = total height in feet
$I = (\pi d^3 t/8) \, (1/12)$, moment of inertia of the section, ft^4
w = weight per foot of the vessel, lb/ft
t = shell thickness, in.
d = shell mean corroded diameter, ft.

Substituting the above values into the equation for T gives

$$T = (2.70/10^5) \, (H/d)^2 \, (wd/t)^{1/2} \text{ sec/cycle}$$

If the vessel operates at higher temperature the period T at the new operating temperature can be found from $T' = T \times (29 \times 10^6/E'$ at operating temperature$)^{1/2}$.

In practice, tall process vessels have either a stepped-down shell thickness or/ and sections of different diameters, representing a system with unevenly distributed mass and flexibility. In such systems, Rayleigh's method is used to determine the first natural period T. Although Rayleigh's method applies only to undamped systems, it yields the fundamental period T with a sufficient accuracy for most engineering problems.

If the vibration of a column is assumed to be undamped harmonic, the sum of the elastic potential energy P.E. and the kinetic energy K.E. remains constant. Maximum kinetic energy K.E. occurs, when the system passes through the equilibrium position and the elastic potential energy is zero. The maximum elastic energy P.E. occurs, when system is at maximum displacement, with zero kinetic energy K.E. Both maximum energies must be equal:

$$\text{Maximum K.E.} = \text{Maximum P.E.}$$

For a simple, one-mass, vibrating system,

$$WV^2/2g = Fy/2$$
$$W(\omega_n y) = Fy/2$$

and

$$\omega_n = (gFy/Wy^2)^{1/2}$$

where

y = amplitude, the maximum deflection of the center of gravity of the mass
V = maximum velocity of the mass, equal to $y\omega_n$ for a simple harmonic motion
ω_n = angular natural undamped frequency of the system, rad/sec
F = initial acting force
W = weight of the mass
g = 32.2 ft/sec^2, gravitational acceleration.

In order to apply the above equations to a cantilever beam, the following simplifying assumptions have to be made.

1. The distributed weights of the beam sections are assumed to be concentrated or "lumped" at centers of gravity of the beam sections with unchanged stiffness along the length of the beam (see Fig. 4.18). This assumption considerably simplifies the computations for the deflections y of the centers of gravity. The greater the number of sections, the higher the final accuracy achieved.

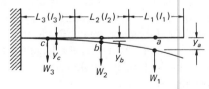

y_a, y_b, y_c = deflections of centers of gravity a, b, c of individual sections
W_1, W_2, W_3 = operating weights of individual sections
I_1, I_2, I_3 = moments of inertia of individual sections

Fig. 4.18.

2. In order to compute the maximum deflections of the centers of gravity of the lumped masses the most probable vibration curve of the system for the first period has to be assumed. The closer the assumed curve to the actual curve, the more accurate the resulting T. However, even if the selected curve differs from the actual to a considerable degree, the computed T is still close enough. Generally, the deflection bending curve of a vessel subjected to its own weight (initial acting force $F = W$) is assumed, and deflections of the centers of gravity $y_a, y_b, y_c, \ldots, y_n$ are computed.

3. Using the above weights and calculated deflections, the natural, undamped, angular frequency ω_n of the system in Fig. 4.18 may be found from the following equations:

$$\omega_n = [g(W_1 y_a + W_2 y_b + \cdots + W_n y_n)/(W_1 y_a^2 + W_2 y_b^2 + \cdots + W_n y_n^2)]^{1/2}$$

or

$$T = 2\pi [\Sigma W y^2 / g \Sigma W y]^{1/2}.$$

Thus calculated T will be slightly (a few percent) lower than the actual T, since the static, bending deflection curve is not actual, dynamic vibrational curve. The signs of the deflections must always be taken as positive.

The weights $W_1, W_2, \ldots, W_n$ used in the computations are the operating weights. To find the deflections $y_a, y_b, \ldots, y_n$ any analytical method can be used. However, if the computations are done by hand the following numerical method is the most convenient and simple to apply.

Computation of the Transverse y Deflections

Basically, this is the conjugate beam method applied to a cantilever beam. (a) The slope of the elastic bending curve of the real beam is equal to the shear at the corresponding point of the conjugate beam (the corresponding fictitious beam with the same length as the real beam but adjusted supports) which has the M/EI area of the real beam as its (elastic) load. (b) The deflection y of the real beam at any point with respect to its original position is equal to the bending moment at the corresponding point of the conjugate beam which has the M/EI area of the real beam as its (elastic) load.

A step-by-step design procedure is outlined below and in Figs. 4.19 and 4.20.

1. The weights of the vessel sections W_1, W_2, W_3 are computed and assumed to act at the centers of gravity (points a, b, and c in Fig. 4.19). For the sake of simplicity it is assumed here that the moment of inertia I and modulus of elasticity E are constant through the entire beam length, so that they can be separated and used in the final formula.

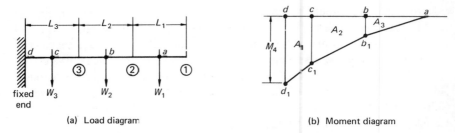

(a) Load diagram   (b) Moment diagram

Fig. 4.19.   Real cantilever beam.

Compute the bending moments at points a, b, and c and at the fixed end due to the concentrated loads W_1, W_2, and W_3:

$$M_a = 0$$
$$M_b = W_1(L_1 + L_2)/2$$
$$M_c = W_1[(L_1/2) + L_2 + (L_3/2)] + [W_2(L_2 + L_3)/2]$$
$$M_d = W_1[(L_1/2) + L_2 + L_3] + W_2[(L_2/2) - L_3] + (W_3 L_3/2).$$

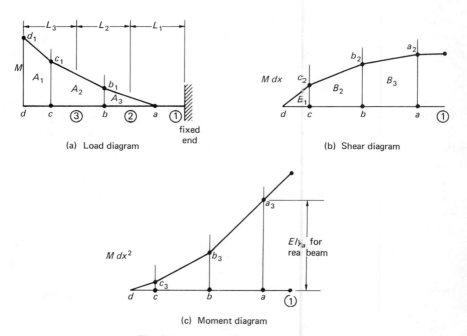

(a) Load diagram   (b) Shear diagram

(c) Moment diagram

Fig. 4.20.   Conjugate cantilever beam.

Sketch the moment diagram to a suitable scale, with $M_b = bb_1$, $M_c = cc_1$, and $M_d = dd_1$. Compute the areas

$$A_1 = (dd_1 + cc_1)(dc)/2$$
$$A_2 = (cc_1 + bb_1)(cb)/2$$
$$A_3 = (bb_1)(ba)/2$$

2. The M-diagram of the real beam becomes the fictitious loading diagram of the corresponding conjugate beam in Fig. 4.20(a). The first numerical integration (using the trapezoidal approximation) gives the shear diagram of the conjugate beam equivalent to the slope diagram of the real beam in Fig. 20(b). The diagram is drawn using the following values:

$$cc_2 = A_1, \quad bb_2 = A_1 + A_2, \quad \text{and} \quad aa_2 = A_1 + A_2 + A_3.$$

3. The second numerical integration yields the moment diagram of the conjugate beam, which is equal to the deflection diagram of the real beam in Fig. 4.20(c). The deflection curve of the real beam falls inside the polygon. Compute areas B_1, B_2, and B_3:

$$B_1 = (cc_2/2)(dc)$$
$$B_2 = (cc_2 + bb_2)(cb)/2$$
$$B_3 = (bb_2 + aa_2)(ba)/2.$$

Sketch the $M\,dx^2$ diagram using the following values:

$$cc_3 = B_1, \quad bb_3 = B_1 + B_2, \quad \text{and} \quad aa_3 = B_1 + B_2 + B_3.$$

4. The deflections can now be found from the equations, for instance,

$$y_a = aa_3/EI.$$

Example 4.4. As an example, the first period of vibration T is computed for the process column used in Example 1.2 in Section 1.6, and shown here in Fig. 4.21 without the reboiler.

The vessel is divided into six sections and the center of gravity of each section is assumed to be at its geometrical centers (points a to f) and to be the loading points of forces equal to the weights W of the respective sections. The moment values are computed at each loading point as well as at each point where the moment of inertia I of the vessel shell changes. In Fig. 4.22 the M/I diagram is

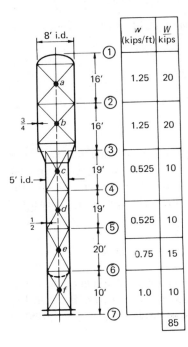

	w (kips/ft)	W kips
①		
16'	1.25	20
②		
16'	1.25	20
③		
19'	0.525	10
④		
19'	0.525	10
⑤		
20'	0.75	15
⑥		
10'	1.0	10
⑦		
		85

Fig. 4.21.

sketched and applied as a fictitious loading diagram to the conjugate beam. Since the moment of inertia I of the shell changes at point 3 a step will appear there in the moment diagram. The subsequent numerical integrations yield the values for computation of the deflections y.

The values of ΣWy and ΣWy^2 are calculated in Table 4.4, and T is calculated as follows:

$$T = 2\pi \left[\frac{7.9483 \times 10^6}{386 \times 663920} \right]^{1/2} = 2\pi(0.17611) = 1.11 \text{ sec/cycle}$$

say $T = 1.15$ sec/cycle.

Table 4.4.

POINT	y (in.)	W (lb)	Wy (lb-in.)	Wy^2 (lb-in.^{2})
a	15.1	20,000	302,000	4.56×10^6
b	11.5	20,000	230,000	2.645×10^6
c	7.58	10,000	75,800	0.575×10^6
d	3.87	10,000	38,700	0.1498×10^6
e	1.11	15,000	16,650	0.0185×10^6
f	0.077	10,000	770	—
		$\Sigma Wy, \Sigma Wy^2$:	663,920	7.9483×10^6

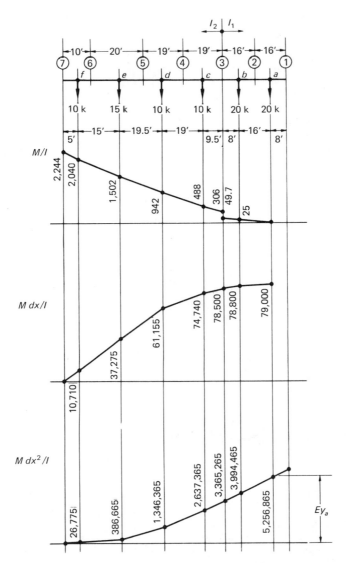

Fig. 4.22.   See page 118 for explanation of figure.

$$I_1 = \pi \times 4{,}0313^3 \times \frac{0.75}{12} = 12.85 \text{ ft}^4$$

$$I_2 = \pi \times 2{,}521^3 \times \frac{0.50}{12} = 2.09 \text{ ft}^4$$

M: $M_b = 20 \times 16 = 320$ kips·ft
$\quad M_3 = 320 + 40 \times 8 = 640$
$\quad M_c = 640 + 40 \times 9.5 = 1{,}020$
$\quad M_d = 1{,}020 + 50 \times 19 = 1{,}970$
$\quad M_e = 1{,}970 + 60 \times 19.5 = 3{,}140$
$\quad M_f = 3{,}140 + 75 \times 15 = 4{,}265$
$\quad M_7 = 4{,}265 + 85 \times 5 = 4{,}690$

M/I: $M_b/I_1 = 25$ kips/ft^3
$\quad M_3/I_1 = 49.7$
$\quad M_3/I_2 = 306$
$\quad M_c/I_2 = 488$
$\quad M_d/I_2 = 942$
$\quad M_e/I_2 = 1502$
$\quad M_f/I_2 = 2040$
$\quad M_7/I_2 = 2244$

$M\,dx/I$: $\dfrac{2{,}244 + 2{,}040}{2} \times 5 = 10{,}710$ kips/ft^2

$\qquad \dfrac{2{,}040 + 1{,}502}{2} \times 15 = \dfrac{26{,}565}{37{,}275}$

$\qquad \dfrac{1{,}502 + 942}{2} \times 19.5 = \dfrac{23{,}880}{61{,}155}$

$\qquad \dfrac{942 + 488}{2} \times 19 = \dfrac{13{,}585}{74{,}740}$

$\qquad \dfrac{488 + 306}{2} \times 9.5 = \dfrac{3{,}760}{78{,}500}$

$\qquad \dfrac{49.7 + 25}{2} \times 8 = \dfrac{300}{78{,}800}$

$\qquad \dfrac{25}{2} \times 16 = \dfrac{200}{79{,}000}$

$M\,dx^2/I$: $\dfrac{10{,}710}{2} \times 5 = \quad 26{,}775$ kips/ft

$\qquad \dfrac{10{,}710 + 37{,}275}{2} \times 15 = \dfrac{359{,}890}{386{,}665}$

$\qquad \dfrac{37{,}275 + 61{,}155}{2} \times 19.5 = \dfrac{959{,}700}{1{,}346{,}365}$

$\qquad \dfrac{61{,}155 + 74{,}740}{2} \times 19 = \dfrac{1{,}291{,}000}{2{,}637{,}365}$

$\qquad \dfrac{74{,}740 + 78{,}500}{2} \times 9.5 = \dfrac{727{,}900}{3{,}365{,}265}$

$\qquad \dfrac{78{,}500 + 78{,}800}{2} \times 8 = \dfrac{629{,}200}{3{,}994{,}465}$

$\qquad \dfrac{78{,}800 + 79{,}000}{2} \times 16 = \dfrac{1{,}262{,}400}{5{,}256{,}865}$

Deflections:

$$y_a = \frac{5.256 \times 10^6 \times 10^3}{29 \times 10^6 \times 144} = 1.258 \text{ ft}$$

$$y_b = \frac{3{,}994 \times 10^6 \times 10^3}{29 \times 10^6 \times 144} = 0.956$$

$$y_c = \frac{2{,}637 \times 10^6 \times 10^3}{29 \times 10^6 \times 144} = 0.631$$

$$y_d = \frac{1.3461}{2.9 \times 1.44} = 0.3223$$

$$y_e = \frac{0.3867}{2.9 \times 1.44} = 0.0926$$

$$y_f = \frac{0.0268}{2.9 \times 1.44} = 0.0064$$

4.8. ILLUSTRATIVE EXAMPLE

Detailed mechanical computations are to be prepared for the process column with the design data given below.

1. Design Data

Shell inside diameter: 5 ft 6 in.
Shell overall length: 160 ft
Support skirt height: 10 ft
Operating hydrostatic head: 10 ft above the bottom t.l.
Operating pressure: 180 psig
Design pressure: 220 psig
Operating temperature: top half, $-45°F$
 bottom, $-10°F$
Design temperature: $-50°F$
Corrosion allowance: $\frac{1}{16}$ in.
Material: shell, SA 516 gr. 55 $S_a = 55{,}000/4 = 13{,}750$ psi, use $S_a = 13{,}700$ psi
 skirt, top SA 516 gr. 55   (Charpy V-notch impact tested)
 bottom SA 285 gr. C
PWHT: yes
X-ray: full
Weld efficiency: 100 percent
Trays: C. S. perforated, number 90, weir height 4 in.
Cold insulation thickness: 5 in.
Flange rating: 150 lb min.
Minimum hydrostatic test pressure: 330 psig
Location: Houston, Texas.

The design and fabrication shall comply with the requirements of the applicable subsections of the ASME Pressure Vessel Code Section VIII, Division 1, 1977 Edition, up to and including 1977 Winter Addenda.

2. Shell Thickness Required for Internal Pressure

Shell thickness:

$$t = PR_i/(SE - 0.6P) = (220 \times 33.06)/(13{,}700 - 132) = 0.536 + \text{C.A.}$$
$$= 0.536 + 0.063 = 0.599 \text{ in.}$$

Use $\frac{5}{8}$-in.-thick plate.

$$P = (13{,}700 \times 0.5625)/(33.06 + 0.6 \times 0.5625) = 230 \text{ psi}$$
$$P_{\text{unc}} = (13{,}700 \times 0.625)/(33.0 + 0.6 \times 0.625) = 256 \text{ psi}$$

2:1 ellipsoidal head thickness, (min.)

$$t_h = PD_i/(2SE - 0.2P) = (220 \times 66.125)/(2 \times 13{,}700 - 44)$$
$$= 0.532 + \text{C.A.} = 0.532 + 0.063 = 0.59 \text{ in.}$$

Use $\frac{5}{8}$=in.-thick plate.

Bottom:

$$t_h = (224 \times 66.125)/(2 \times 13{,}700 - 44) = 0.542 + 0.063 = 0.605 \text{ in.}$$

For bottom head use the same thickness as the bottom shell section: $1\frac{3}{16}$ in. thick, as computed from the wind loads.

3. Shell Thickness Based on Combined Wind, Weight and Pressure Loads

Effective shell diameter D_e:

$$D_e = (\text{o.d. column} + 10 \text{ in. ins.}) + (\text{o.d. 6-in. pipe} + \text{ins.}) + \text{platform} + \text{ladder}$$
$$= (67.25 + 10) + (6.625 + 6) + 6 + 12 = 107.9 \text{ in.} \doteq 9 \text{ ft.}$$

Basic wind pressure p = 35 psf. (See remarks at end of example.)

HEIGHT ZONE	WIND PRESSURE (psf)	$w = B \times D_e \times p_z$ (lb/ft OF HEIGHT)
0' to 30'	25	$w = 0.6 \times 9 \times 25 = 135$
30' to 50'	35	$w = 0.6 \times 9 \times 35 = 189$
50' to 100'	45	$w = 0.6 \times 9 \times 45 = 243$
100' to 170'	55	$w = 0.6 \times 9 \times 55 = 297$

Wind loads

$$l_w = (4M \times 12)/(\pi D^2 \times 144) \quad = 0.106\, M/D^2 \text{ lb/in.}$$

At section 2: $l_w = (0.106 \times 732{,}650)/5.5^2 \quad = \quad 2{,}570 \text{ lb/in.}$
At section 3: $l_w = (0.106 \times 2{,}075{,}900)/5.5^2 = \quad 7{,}274$
At section 4: $l_w = (0.106 \times 2{,}772{,}500)/5.5^2 = \quad 9{,}715$
At section 5: $l_w = (0.106 \times 3{,}533{,}900)/5.5^2 = 12{,}383$
At section 6: $l_w = (0.106 \times 3{,}934{,}850)/5.5^2 = 13{,}790.$

Weight loads

$$l_d = W/12\pi D \text{ lb/in.}$$

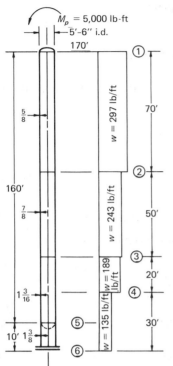

Computation of Wind Loads and Moments		
Wind load (lb)	Shear (lb)	Moment (lb–ft)
		$M_p = 5000$
297 × 70 = 20,790		
	20,790	732,650
12,150		
	32,940	2,075,900
3,780		
	36,720	2,772,500
2,700		
	39,420	3,533,900
1,350		
	40,770	3,934,850

Moment at base $M_b = 3,934,850$ lb-ft

Fig. 4.23.

At section 2:	$l_d = 94,850/(12 \times \pi \times 5.5)$	=	457 lb/in.
At section 3:	$l_d = 164,900/(12 \times \pi \times 5.5)$	=	795
At section 5:	$l_d = 246,850/(12 \times \pi \times 5.5)$	=	1,190
At section 6:	$l_d = 278,750/(12 \times \pi \times 5.5)$	=	1,342.

Shell thickness computations

At section 2:
  (a) Windward side:

$$l_w - l_d + PR/2 = 2,570 - 457 + (220 \times 33.325)/2 = 5,785 \text{ lb/in.}$$

required shell thickness:

$$t = 5,785/13,700 + \text{C.A.} = 0.422 + 0.0625 = 0.489 \text{ in.}$$

Hoop pressure thickness governs, use $\frac{5}{8}$-in. thick plate.
(b) Leeward side:

$$l_w + l_d = 2,570 + 457 = 3,027 \text{ lb/in.}$$

Code allowable stress $B = 13,700$ psi in compression (buckling). Actual stress in compression:

$$S_c = 3,027/0.5625 = 5,381 \text{ psi.}$$

At section 3:
(a) Windward side:

$$l_w - l_d + (PR/2) = 7,274 - 795 + 3,670 = 10,149 \text{ lb/in.}$$

required $t = 10,149/13,700 + 0.0625 = 0.740 + 0.0625 = 0.803$ in.

Use $\frac{7}{8}$-in.-thick plate.
(b) Leeward side:

$$l_w + l_d = 7,274 + 795 = 8,069 \text{ lb/in.}$$

Code allowable stress $B = 13,700$ psi. Actual stress:

$$S_c = 8,069/0.8125 = 9,950 \text{ psi.}$$

At section 5:
(a) Windward side:

$$l_w - l_d + (PR/2) = 12,383 - 1,190 + 3,670 = 14,863 \text{ lb/in.}$$

required $t = 14,863/13,700 + 0.0625 = 1.085 + 0.0625 = 1.147$ in.

Use $1\frac{3}{16}$-in.-thick plate.
(b) Leeward side:

$$l_w + l_d = 12,383 + 1,190 = 13,573 \text{ lb/in.}$$

Code allowable stress, $B = 13,700$ psi. Actual stress:

$$S_c = 13,573/1.125 = 12,064 \text{ psi}$$

4. Support Skirt Design

(a) Skirt thickness at section 5, based on the attachment weld:

$$t_{sk} = (l_w + l_d)/ES_a = 13{,}573/(0.8 \times 13{,}700) = 1.24 \text{ in.}$$

Use $1\frac{3}{8}$-in.-thick plate.

(b) Anchor bolts:
   At skirt base 6 flare the skirt to 11 ft to accommodate anchor bolt number.

Tension:

$$l_w - l_d = (3{,}934{,}850 \times 4)/(\pi \times 11.5^2 \times 12)] - [278{,}750/(\pi \times 11.5 \times 12)]$$
$$= 3{,}158 - 643 = 2{,}515 \text{ lb/in.}$$

Required bolt area:

$$A_b = (\pi \times \text{B.C.} \times 2{,}515)/(N \times S_a)$$
$$= (\pi \times 11.5 \times 12 \times 2{,}515)/(20 \times 15{,}000) = 3.62 \text{ in.}^2$$

Use 20 $2\frac{1}{2}$-in. bolts with the stress area 4.00 in.^{2}

$$\text{spacing} = (11.5 \times \pi)/20 = 21.8 \text{ in.}$$

Maximum stress in bolt under empty condition ($W_e = 134{,}800$ lb.):

$$S = (3{,}158 - 134{,}800/\pi \times 11.5 \times 12) \pi \times \text{B.C.}/N \times A_b$$
$$= 15{,}450 \text{ psi} < 18{,}000 \text{ psi}$$

(c) Base details:
   Base ring thickness: material A285 gr. C, $S_a = 16{,}700$ psi no wind, 22,200 psi with wind.

Wind + operating weight:

$$l_w + l_d = (3{,}158 \times 11.5^2/11^2) + (643 \times 11.5/11) = 4{,}122 \text{ lb/in.}$$

Bearing pressure:

$$p = (l_w + l_d)/b = 4{,}122/9.5 = 0.435 \text{ kips}$$
$$t_b = (0.15 \times 0.435 \times 6.125^2 \times 20/22.2)^{1/2} = 1.48 \text{ in.}$$

Use $1\frac{1}{2}$-in.-thick plate.

Weight only:

$$t_b = (0.15 \times 0.071 \times 6.125^2 \times 20/16.7)^{1/2} = 0.67 \text{ in.}$$

Skirt-to-base weld:

$$\text{leg size } w = (l_w + l_d)/f_w = 4{,}122/(2 \times 1.33 \times 13\,700 \times 0.55) = 0.21 \text{ in.}$$

Use $\frac{5}{16}$ -in. fillet weld all around on both sides.
   Top ring thickness:

$$t_r = [(4.13 \times 15{,}000 \times 4 \times 1.25)/(4 \times 20{,}000 \times 3.5)]^{1/2} = 1.05 \text{ in.}$$

Use 1 in. thick plate.
   All other dimensions taken from Table A3.1 at the back of this book.

(d) Stresses at sharp cone–cylinder intersection of skirt support shown in Fig.
4.24 (for stress formulas and their derivation see Section 8.6):
   At the cone–cylinder intersection:

$$D = 69.75 \text{ in.}$$
$$\alpha = 15 \text{ degrees}$$
$$n = 1$$

Windward side, tension:

$$l = l_w - l_d = 12{,}383 - 1{,}190 = 11{,}193 \text{ lb/in.}$$
$$P_e = (4l/D) = 642 \text{ psi}$$

Leeward side, compression:

$$l = l_w + l_d = -13{,}573 \text{ lb/in.}$$
$$P_e = -778 \text{ psi}$$

Stresses in cylindrical shell:

$$
\begin{aligned}
\sigma_L &= (P_e R/t)\,(0.5 \pm X\sqrt{R/t}) \\
&= (-778 \times 34.87/1.375)\,(0.5 \pm 0.155\sqrt{34.87/1.375}) \\
&= -25{,}254 \text{ psi} \quad \text{and} \quad +5{,}524 \text{ psi}
\end{aligned}
$$

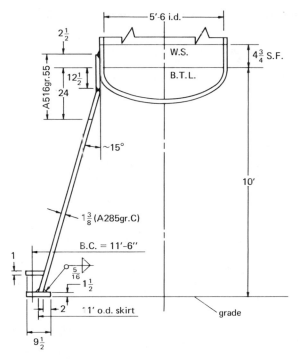

Fig. 4.24. Support skirt detail.

$$\sigma_{tm} = P_e Y (R/t)^{3/2} = -778 \times 0.088 \,(34.87/1.375)^{3/2} = -8743 \text{ psi} < 1.33 S_a$$

$$\text{or} \quad -8{,}743 - (25{,}254/2) = -21{,}370 \text{ psi} < 2 S_a$$

Stresses in conical shell:

$$\sigma_L = (P_e R/t)\,[(0.5/n \cos \alpha) \pm U \sqrt{R/t}\,)$$

$$= (-778 \times (34.87/1.375)\,[(0.5/1 \times 0.964) \pm 0.155 \sqrt{34.87/1.375}\,]$$

$$= -25{,}629 \text{ psi} \quad \text{and} \quad -5{,}155 \text{ psi}$$

$$\sigma_{tm} = -8{,}743 \text{ psi} < 1.33 S_a$$

$$\text{or} \quad -8{,}743 - 25{,}629/2 = -21{,}560 \text{ psi} < 2 S_a$$

5. Deflection Due to Wind—Corroded Condition (Fig. 4.25)

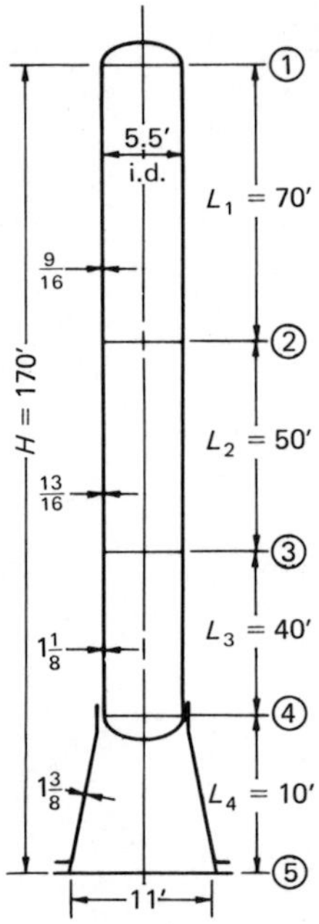

	Wind load W (lb)	Wind shear Q (lb)	Wind moment M (lb-ft)	Moment of inertia I (ft^4)	Remarks
①			$M_1 = 5,000$		
$L_1 = 70'$	20,790			$I = \dfrac{\pi r^3 t}{3.17}$	$E = 29 \times 10^6$ psi $= 4.176 \times 10^9$ psf
②		20,790	0.733×10^6		
$L_2 = 50'$	12,150			4.63	
③		32,940	2.076×10^6		
$L_3 = 40'$	6,480			6.48	
④		39,420	3.53×10^6		
$L_4 = 10'$	1,350			equivalent $I = \pi(8.8)^3 \dfrac{1.375}{12}$	
⑤		40,770	3.935×10^6	$= 245.31$	

Fig. 4.25.

$$y_1 = \frac{70^2}{4.176 \times 10^9 \times 3.19}\left[\frac{2.079 \times 10^4 \times 70}{8} + \frac{5 \times 10^3}{2}\right] = 0.0678 \text{ ft}$$

$$y_2 = \frac{5^2 \times 10^2}{4.176 \times 10^9 \times 4.63}\left[\frac{2.079 \times 10^4 \times 50}{3} + \frac{1.215 \times 10^4 \times 50}{8} + \frac{7.33 \times 10^5}{2}\right] = 0.1021$$

$$y_3 = \frac{4^2 \times 10^2}{4.176 \times 10^9 \times 6.98}\left[\frac{3.29 \times 10^4 \times 40}{3} + \frac{6.48 \times 10^3 \times 40}{8} + \frac{2.076 \times 10^6}{2}\right] = 0.0892$$

$$y_4 = \frac{10^2}{4.176 \times 10^9 \times 245.31}\left[\frac{3.9 \times 10^4 \times 10}{3} + \frac{1.35 \times 10^3 \times 10}{8} + \frac{3.53 \times 10^6}{2}\right] = 0.0002$$

$$\Delta_{1-2} = \frac{70 \times 50}{4.176 \times 10^9 \times 4.63}\left[\frac{2.079 \times 10^4 \times 50}{2} + \frac{1.215 \times 10^4 \times 50}{6} + 7.33 \times 10^5\right] = 0.2451$$

$$\Delta_{1-3} = \frac{120 \times 40}{4.176 \times 10^9 \times 6.48}\left[\frac{3.29 \times 10^4 \times 10}{2} + \frac{6.48 \times 10^3 \times 40}{6} + 2.076 \times 10^6\right] = 0.4926$$

$$\Delta_{1-4} = \frac{160 \times 10}{4.176 \times 10^9 \times 245.31}\left[\frac{3.94 \times 10^4 \times 10}{2} + \frac{1.35 \times 10^5 \times 10}{6} + 3.53 \times 10^6\right] = \frac{0.0058}{1.0022 \text{ ft}}$$

Maximum deflection at the top:

$$y = 12.03 \text{ in.}$$

$$12.03/1.70 = 7.07 \text{ in.}$$

per 100 ft of height in corroded condition (approved by client).

6. Fundamental Period of Vibration—Uncorroded Condition (Fig. 4.26)

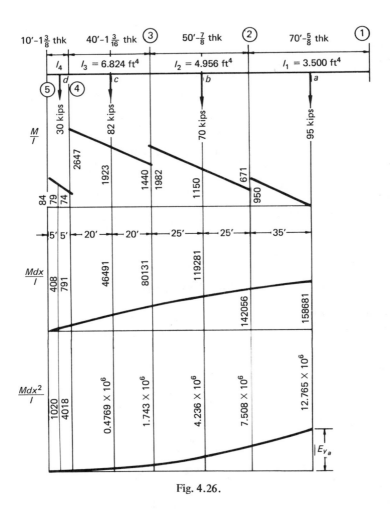

Fig. 4.26.

Deflections:

$$y_a = \frac{12.765 \times 10^6 \times 10^3}{29 \times 10^6 \times 144} = 3.056 \text{ ft}$$

$$y_b = \frac{4.236 \times 10^6 \times 10^3}{29 \times 10^6 \times 144} = 1.014$$

$$y_c = \frac{0.4769 \times 10^6 \times 10^3}{29 \times 10^6 \times 144} = 0.114$$

$$y_d = \frac{1020 \times 10^3}{29 \times 10^6 \times 144} = 0.030$$

POINT	y (in.)	W (kips)	Wy	Wy^2
a	36.68	95	3,485	127,815
b	12.17	70	852	10,368
c	1.37	82	112	154
d	0.03	30	0.9	—
		$\Sigma Wy, \Sigma Wy^2$:	4,449.9	138,337

Period:

$$T = 2\pi(138,337/386 \times 4,449)^{1/2} = 1.783 \text{ sec/cycle}$$

Say $T = 1.85$ sec/cycle.

7. Investigation of Possible Wind-caused Vibrations

First wind critical velocity:

$$V_1 = 3.40d/\text{T} = 3.40 \times 6.38/1.85 \doteq 12.0 \text{ mph.}$$

Second critical velocity:

$$V_2 = 6.25 V_1 = 6.25 \times 12 = 75 \text{ mph} > 60 \text{ mph.}$$

Equivalent load at top:

$$F = 0.00086 \times 35 \times 6.38 \times 170 \times 12^2 = 4,685 \text{ lb.}$$

Moment at bottom T.L.:

$$F \times 160 = 752,175 \text{ lb-ft.}$$

Wind moment as computed is higher, no further stress check is required. Estimated cyclic lifetime:

Longitudinal cyclic stress at BTL: $0.106 \times 752{,}175/5.5^2 = 2{,}635$ lb/in.
Stress range $S_r = 2 \times 2{,}635/1.125 = 4{,}685$ psi.
Stress concentration factor $\beta = 2.0$.
Number of cycles to failure: $N = (780{,}000/2 \times 4{,}685)^5 = 4.06 \times 10^9$ cycles.
S.F. = 20; $N/20 = 2.03 \times 10^8$.
Safe vibration time: $(2.03 \times 10^8 \times 1.8/3{,}600) \times (1/24 \times 356) = 12$ years.

8. Field Hydrotest in Corroded Condition

Maximum stress at any point in the vessel should not exceed $0.8\ S_y$ during the hydrotest.

Maximum allowable: $S_a = 13{,}700$ psi
Yield strength: $S_y = 30{,}000$ psi

Maximum allowable ratio of hydrotest stress/design stress:

$$\frac{0.8 \times 30{,}000}{13{,}700} \leqslant 1.75.$$

Pressures in the vertical column during hydrotest (see Fig. 4.27):

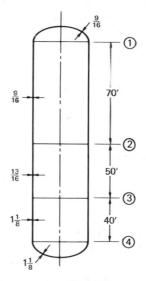

Fig. 4.27.

Minimum test pressure at top: $1.5 \times 220 =$ 330 psig

Liquid static head to section 2: $71.4 \times 0.434 \doteq$ 31

             Test pressure at 2   361

Liquid static head from 2 to 3: $50 \times 0.434 \doteq$ 22

             Test pressure at 3   383

Liquid static head from 3 to 4: $40 \times 0.434 \doteq$ 18

             Test pressure at 4   401

Liquid static head from 4 to bottom: $1.4 \times 0.434 \doteq$ 1

             Test at bottom   402

Maximum allowable test pressure for 150-lb flanges at bottom: 425 psig $>$ 402 psig.

Pressure stress ratios for shell:

$$RT = \frac{P(R_i + 0.6t)/t}{S_a E} = \frac{P_H(R_i + 0.6t)}{S_a Et} \leqslant 1.75$$

At section 2: $RT_2 = \dfrac{361\,(33.063 + 0.6 \times 0.5625)}{13{,}700 \times 1.0 \times 0.5625} = 1.57 < 1.75$

At section 3: $RT_3 = \dfrac{383\,(33.063 + 0.6 \times 0.8125)}{13{,}700 \times 1.0 \times 0.8125} = 1.152 < 1.75$

At section 4: $RT_4 = \dfrac{401\,(33.063 + 0.6 \times 1.125)}{13{,}700 \times 1.0 \times 1.125} = 0.876 < 1.75$

For head: $RT = \dfrac{P(D_i + 0.2t)}{2SEt}$

At bottom: $RT = \dfrac{402\,(66.125 + 0.6 \times 1.125)}{2 \times 13{,}700 \times 1.0 \times 1.125} = 0.87 < 1.75$

9. Approximate Weights for Design

(1) *Operating weight W_o:*

To section 2	
Top head, $\frac{5}{8}$ in. thick, 2:1 ellipsoidal	1,200 lb
Shell, $\frac{5}{8}$ in. thick, 472 lb/ft $\times$ 70 ft	33,000
Top demister with support	200
Tray supports, 2 in. $\times$ $\frac{1}{4}$ in. thick, 35 lb/ring $\times$ 45	1,600
Manholes, 2, 20-in., 750 lb $\times$ 2	1,500

Nozzles (estimated)	500
Clips for platforms and ladder (estimated)	500
Insulation, cold, 5 in. thick, 90 lb/ft × 70 ft	(6,300)
Platforms 3, 3 × 25 ft² × 35 lb/ft²	(2,650)
Piping (estimated)	(5,000)
Ladder 20 lb/ft × 70 ft	(1,400)
Operating liquid on trays, 1,482 lb/ft × 45 × (4 + 1)/12	(27,000)
Trays, 310 lb/tray × 45	(14,000)
W_o at section 2	94,850 lb

Shell, $\frac{7}{8}$ in. thick, 660 lb/ft × 50 ft	33,000
Tray support rings, 25 × 35 lb/ring	900
Manholes, 2, 20-in., 800 lb × 2	1,600
Nozzles (estimated)	500
Clips (estimated)	500
Insulation, 50 ft × 90 lb/ft	(4,500)
Platforms, 2 × 25 ft² × 35 lb/ft²	(1,750)
Piping (estimated)	(3,000)
Ladder, 20 lb/ft × 50 ft	(1,000)
Operating liquid, 1,482 lb/ft × $(\frac{5}{12})$ × 25	(15,500)
Trays, 310 lb/tray × 25	(7,800)
W_o at section 3	164,900 lb

Shell, $1\frac{3}{16}$ in. thick, 900 lb/ft × 40 ft	36,000
Tray supports, 35 lb/ring × 20	700
Manholes, 2, 20-in., 850 lb × 2	1,700
Nozzles (estimated)	500
Clips (estimated)	500
Insulation, 90 lb/ft × 40 ft	(3,600)
Platforms, 2 × 25 ft² × 35 lb/ft²	(1,750)
Piping (estimated)	(3,000)
Ladder, 50 ft × 20 lb/ft	(1,000)
Operating liquid, 1,482 lb/ft × $[(\frac{5}{12}) × 20 + 10]$	(27,000)
Trays, 310 lb/tray × 20	(6,200)
W_o at section 5	246,850 lb

Support skirt, flared to 11 ft, $1\frac{3}{8}$ in. thick	15,000
Base ring, top ring, stiffeners	3,000
Fireproofing, 4 in. thick	(10,000)
Operating liquid in bottom head	(1,500)
Bottom head, 2:1 ellipsoidal, $1\frac{3}{16}$ in. thick	2,400
W_o at base 6	278,750 lb

(2) *Erection weight:*

$$W_e = 134{,}800 \text{ lb}, \quad \text{no trays}$$
$$W_e = 162{,}800 \text{ lb}, \quad \text{with trays.}$$

(3) *Shop test weight:*

$$W_T = 134{,}800 + 162 \text{ ft} \times 1{,}482 \text{ lb/ft} = 375{,}000 \text{ lb.}$$

(4) *Field test weight:*

$$W_T = 278{,}750 \text{ lb} - \text{ins.} \ 14{,}400 \text{ lb} + 122 \times 1{,}432 \text{ lb/ft} = 445{,}150 \text{ lb}$$

10. Remarks

1. When computations for shell thickness were first computerized, the tendency was to step up the shell wall by $\frac{1}{16}$ in. to save on material as much as possible. The result was too many shell sections with different thicknesses. This did not prove to be most economical and practical for fabrication and deflection computations. Usually, *to divide a tall column into three to five sections has proved to be satisfactory.*

2. For simplicity the standard ASA A58.1-1955 was used in computing the unit wind load in Fig. 4.23. The unit load w can be computed by the standard ASA A58.1-1972 without any effect on the computation procedure of the shell thickness.

3. These computations cover the general features of the design. It is expected that careful attention will be given to the preparation of the weld and other details, as well as to materials quality control, fabrication, inspection, and tests.

5

Supports for Short Vertical Vessels

5.1. SUPPORT LEGS

General Considerations

Small and medium-sized vertical vessels, located on the ground and limited to the dimensions given in Table 5.1 are usually supported on uniformly spaced columns called support legs. If a short vessel is located above ground on structural steel construction or if connected by piping to a reciprocating machine (piston compressor), it is usually supported on a skirt to avoid any vibration problems. To allow good access under the vessel, even for larger-diameter vessels the number of the support legs is held to four, braced against the wind, unless a larger number of legs cannot be avoided.

The structural shapes used for support legs are equal leg angles and I-shapes. The two different ways to weld the angle supports to the vessel shell are shown in Fig. 5.1. The position in (a) offers a greater moment of inertia in resisting the external loads on the vessel. However, the angle leg has to be adjusted to the shell curvature for welding. For larger and heavier vessels, I-shapes are used for support legs. Again there are two possible ways to weld them to shell as shown in Fig. 5.2. The I-shapes in (b) are easier to weld to the shell, but support legs welded as in (a) can carry much heavier eccentric external loads. Their required

Table 5.1.

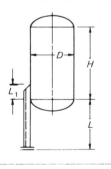

	PRESSURE VESSELS	STORAGE TANKS
Maximum D	6'0"	12'0"
Maximum H/D	5	5
Maximum L/D	2	as required
Number of legs: $N = 3$ for $D \leqslant 3'6"$ $N = 4$ for $D > 3'6"$ $N = 6$ or 8 if required		
Maximum operating temperature = 650°F		

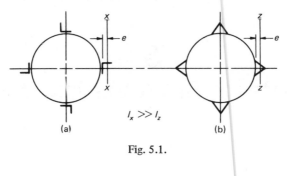

$$I_x \gg I_z$$

Fig. 5.1.

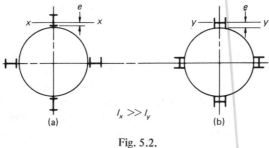

$$I_x \gg I_y$$

Fig. 5.2.

adjusting of the flange to the shell contour can be minimized by using shapes with narrow flanges. Occasionally steel pipes are used for large storage tanks, as shown in Fig. 5.3. Round pipe is particularly suitable for a column, since it possesses a high radius of gyration in all directions and has good torsional resistance. The inside of a column pipe is usually left unpainted, since the seal welds on both ends will be adequate for protection against atmospheric dampness and corrosion. Centroidal axes of pipes used as legs coincide with the center-

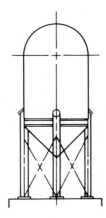

Fig. 5.3.

line of the vessel shell. This eliminates the eccentricity e in the column and base-plate computations.

The immediate task of the designer is to determine by a stress analysis the required dimensions of the following parts:

1. support-leg columns,
2. base plate,
3. leg-to-shell weld size,
4. leg-to-base plate weld size,
5. stresses in the vessel shell at supports,
6. size of anchor bolts.

Support-Leg Columns

The loads imposed on support legs are vertical and horizontal, due to weight and wind or seismic forces. The wind force P_w is computed by multiplying the vertical projected exposed area of the vessel by the wind pressure for the location times the shape factor. The minimum wind pressure is usually taken as 20 psf, which corresponds approximately to a wind velocity of 100 mph. If the vessel is located in a region subjected to earthquakes, the seismic force P_e acting on the vessel will be equal to the weight of the vessel times the earthquake coefficient c described in Chapter 1.

The computed wind load is horizontal and is assumed to be acting in the centroid of the projected exposed surface. The earthquake load P_e also acts horizontally on the center of gravity of the vessel (see Fig. 5.4).

Generally, there are no additional moments from piping or other equipment to be considered. The tops of the support legs are assumed to be welded to a rigid vessel wall that is actually flexible. The anchor bolts are initially pretightened, and as long as some compression between the base plates and the foundation exists due to weight and the initial bolt load, the vessel will have a tendency to overturn about the axis A-A (Fig. 5.4) as a neutral axis and the reactions in the columns due to the overturning moment M_b will be proportional to the distance from the axis A-A.

The vertical reactions C in compression and T in tension at the support base consist of (a) vessel weight, assumed to be equal to W/N, and (b) reaction to the overturning moment M_b.

Using Table 5.2, the following results can be derived. The maximum total axial load at the leeward side (compression) is

$$C_o = (W_o/N) + (4M_b/ND_b) \qquad \text{operating conditions}$$

$$C_T = W_T/N \qquad \text{test conditions.}$$

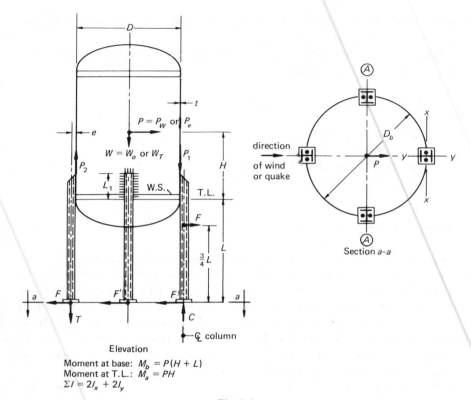

Moment at base: $M_b = P(H + L)$
Moment at T.L.: $M_a = PH$
$\Sigma I = 2I_x + 2I_y$

Fig. 5.4.

Table 5.2. Maximum Vertical Reaction R on Support Column Due to Overturning Moment M about Axis A-A.

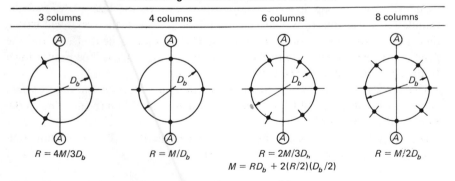

3 columns	4 columns	6 columns	8 columns
$R = 4M/3D_b$	$R = M/D_b$	$R = 2M/3D_b$	$R = M/2D_b$
		$M = RD_b + 2(R/2)(D_b/2)$	

The above results can be expressed by the general formula $R = 4M/ND_b$.

The maximum axial load on the windward side (uplift) is

$$T_o = -(W_o/N) + (4M_b/ND_b) \quad \text{operating conditions}$$
$$T_e = -(W_e/N) + (4M_b/ND_b) \quad \text{empty.}$$

The eccentric loads P_1 and P_2 at column top are

$$P_1 = (W_o/N) + (4M_a/ND) \quad \text{operating conditions}$$
$$P_1 = W_T/N \quad \text{test conditions}$$
$$P_2 = 4(M_a/ND) - (W_o/N) \quad \text{operating conditions}$$
$$P_2 = 4(M_a/ND) - (W_e/N) \quad \text{empty.}$$

For computing the lateral load F per column from the force P, the deflections at the base of all columns under load P are assumed approximately equal to

$$F = PI/\Sigma I$$

where I is the moment of inertia of the column cross section about the axis perpendicular to the direction of the wind or earthquake and ΣI is the summation of the moments of inertia of all column cross sections about the axes perpendicular to the direction of the wind or quake.

To determine the maximum stress in column the column size has to be first selected. Because of number of variables a direct choice of the size is not feasible and successive trials must be made to select the most economical and safe section.

Combined Column Stress in Compression.
(a) Operating conditions:

Axial compression: $\qquad f_a = C/A,$

where A is the cross-section area of the shape.

Bending: $\qquad f_b = [P_1 e/(I_x/c)] + [F(\tfrac{3}{4}L)/(I_x/c)],$

where I_x/c is the section modulus resisting bending of the selected shape. The length $\tfrac{3}{4}L$ was arbitrarily selected to reflect the partial restraining influence at the leg base.

(b) Test conditions:

Axial compression: $f_a = W_T/NA.$

Bending: $f_b = W_T e/N(I_x/c).$

Maximum Compressive Stress in the Column:

$$f = f_a + f_b.$$

Since the allowable unit stress F_a in column for axial loads only is not equal to the allowable unit stress F_b in bending in the absence of an axial load, the question now arises to what value should the total combined compressive stress in column f be limited. To require that f should be no larger than the smaller of the allowable values of F_a or F_b would certainly be conservative and easy to apply. Specifications of some engineering companies do make this requirement, the smaller allowable value usually being the axial allowable stress F_a.

However, a more widely used procedure is the straight-line interaction method used in AISC Handbook [34]. This method of determining the limit stress, without determining the principal stresses, is based on the assumption that if a certain percentage of the strength of a member has been used for axial compression load, the remaining percentage may be used for bending:

$$[(f_a/F_a) + (f_b/F_b)] \leqslant 1 \quad \text{for} \quad f_a/F_a \leqslant 0.15$$

$$\{(f_a/F_a) + [C_m f_b/(1 - f_a/F'_e) F_b]\} \leqslant 1 \quad \text{for} \quad f_a/F_a > 0.15.$$

For good design the expressions on the left should be close to one. The value of F_b can be taken as $0.6F_y$, the yield strength; the reduction factor C_m conservatively equals one. The allowable stress F_a as well as F'_e can be obtained directly from the AISC Manual; they depend on the slenderness ratio kL/r and kL/r_b and the yield strength F_y. The allowable unit stresses F_a, F_b, and F'_e can be increased by one-third if the calculated stresses f_a and f_b are computed on the basis of wind or seismic loads in combination with dead loads. The end conditions of the support legs, which specify the effective length factor k, have to be evaluated by the designer and are subject to differences of opinion. The bottom end of the leg is flat-ended and bearing against the foundation. The anchor bolts are rarely designed to be moment resistant. Theoretically, such a flat end is equivalent to a fixed end, unless it kicks out and bears only on one edge instead of on the whole surface. In practice, a flat-ended column is less restrained at the end than required for full fixity. The support leg can therefore be approximated by a column with end conditions between one end guided, the other at the base fixed ($k = 1$) and one end guided, the other end hinged ($k = 2$). The value of $k = 1.5$ is recommended here. If the translation of the top is prevented by diagonal bracing, both ends are partially rotationally restrained and a smaller value for k could be assumed.

Base Plates

Base plates are primarily used to distribute large, concentrated vertical loads into the concrete foundation so that the concrete is not overstressed and to accomodate the anchor bolts. They are shop welded to the support legs.

The loads used in connection with angle support legs are too low and practical dimensions for base plates are usually selected. I-shape columns are used for larger loads, and the stress distribution between the column and the base plate and between the base plate and the concrete pedestal cannot be accurately determined analytically. However, under an axial column load alone the bearing pressure between the base plate and the concrete base is assumed to be uniformly distributed. The total downward load on the foundation is the sum of the initial tensile load NF_i in the anchor bolts and the column loads. The size of the anchor bolts is usually not to large, and the effect of F_i is neglected. The maximum compressive unit pressure between the base plate and concrete is

$$p = (C/ab) \pm P_1 (d/2)/(a^2 b/6).$$

This equation can only be used if $C/ab \geqslant P_1(d/2)/(a^2 b/6)$, since a tensile stress cannot develop between the base plate and the concrete. This is true in most practical cases, but the difference is small enough to make possible a base plate where the anchor bolts are not required to resist the moment $P_1(d/2)$, (see Fig. 5.5 and ref. 108). Dimensions a and b in Fig. 5.5 for the base plate are first selected to suit the column shape.

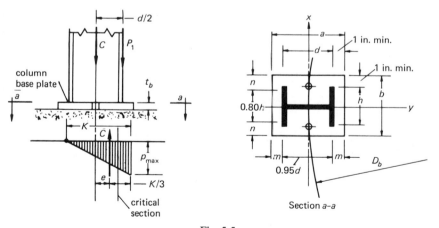

Fig. 5.5.

The load C and the moment $P_1(d/2)$ are assumed to be counteracted by the base reacting moment Ce:

$$Ce = P_1(d/2) \quad \text{or} \quad e = P_1 d/2C.$$

The required K dimension in Fig. 5.5 is

$$K = 3[(a/2) - e]$$

and from

$$K = C/(pb/2)$$

the bearing pressure is

$$p_{max} = 2C/Kb \leqslant F_b = 0.25f'_c,$$

where F_b is the maximum allowable bearing pressure on concrete, per AISC Manual Section 1.5.5. Otherwise the dimensions a and b have to be increased.

Base-Plate Thickness. To determine the base plate thickness t the AISC method is usually used [34]. First, the dimensions m and n in Fig. 5.5 are computed:

$$m = \tfrac{1}{2}(a - 0.95d)$$
$$n = \tfrac{1}{2}(b - 0.80h).$$

Second, the projected portions m and n are considered as uniformly loaded cantilever beams. Using the dimension m or n, depending on the position of the I-shape, the column base plate thickness is computed as follows:

$$S_b = M/Z = (pm^2/2)/(t_b^2/6) = (3pm^2)/t_b^2$$
$$t_b = m \text{ (or } n) \times (3p/S_b)^{1/2} \text{ in.}$$

where S_b is allowable bending stress for the plate material (for carbon structural steel $S_b = 0.6$ times the yield strength), and

$$p = C(2K - m)/bK^2$$

is the average bearing pressure acting across the projected length of m.

The computation assumes that the base plate is in full contact with the concrete pedestal. Base plates computed in this manner are usually satisfactory for the uplift condition. Minimum base plate thickness is not less than 0.5 in.

Leg-to-Shell Connecting Weld Size

Under operating conditions in Fig. 5.4, the shear in the weld is

$$f_s = P_1/(2L_1 + h).$$

The bending stress is

$$f_b = [(Cd/2) + (FL/4)]/Z_w,$$

where Z_w is the linear section modulus of the weld (see Table 10.3). The total combined stress in the weld is

$$f = (f_s^2 + f_b^2)^{1/2}.$$

The size of the weld leg is

$$w = f/f_w,$$

where f_w is the allowable unit force for fillet welds.

Base-Plate Attachment Weld

The column base plate and the column with milled end are in contact when welded and the vertical force and moment are directly transferred from the column into the base plate. If the column end is not milled the connecting weld must transfer the entire column load into the base plate.

The uplift forces T_o and T_e and the shear force F are usually not large enough to govern the weld size and the minimum size of $\frac{1}{4}$ in. continuous or intermittent weld is satisfactory.

Stresses in the Shell

Unfortunately, an exact, workable, analytical solution of the local stresses at the leg-to-shell junction is not available at present time. The support legs for short vertical drums are usually welded to the shell in the region reinforced by the end closure. In addition to the general longitudinal stress in the shell, high localized stresses are imposed on the shell through the welded joint. An estimate of the average maximum stress in the shell can be made as follows.

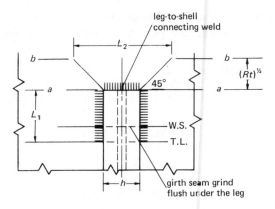

Fig. 5.6.

(a) The approximate maximum general longitudinal stress in shell at section *a-a* in Fig. 5.6. is

$$S_L = (4M_a/\pi D^2 tE) + (PD/4tE) - (W/t\tau DE) \quad \text{in tension,}$$

where E is the shell weld joint efficiency and P is operating pressure, or

$$S_L = (4M_a/\pi D^2 t) + (W/t\pi D) \quad \text{in compression.}$$

(b) The maximum localized stress to cause buckling above the leg top at section *b-b* is

$$S_c = P_1/L_2 t$$

$\leqslant$ maximum Code allowable stress for axially loaded cylindrical shells
    in compression

where

$$L_2 = h + 2(Rt)^{1/2}$$

is the effective resisting length.

(c) The moment $[(FL/4) + (Cd/2)]$ can be used for determining the stresses in shell by method discussed in Chapter 7 or in ref. 105.

If the resulting stresses in the shell are too high, a reinforcement by rings or reinforcing pads must be provided. Increased thickness of the bottom shell section is often sufficient to decrease the stresses.

Size of Anchor Bolts

Anchor bolts are designed to resist the uplift forces and to secure the legs in position. The shear load is small and is resisted by friction. The required bolt tension area A_b per leg is given by

$$A_b = (1/NS_a)[(4M_b/D_b) - W]\text{ in.}^2$$

where

W = either operating weight W_o or empty weight W_e, lb,
S_a = allowable tensile stress under operating or empty conditions for bolt,
N = number of leg supports.

If $W > 4M_b/D_b$ no uplift exists and the minimum practical bolt size, usually $\frac{3}{4}$-1 in., can be selected.

5.2. SUPPORT LUGS

Support Lugs without Stiffening Rings

Support lugs (Fig. 5.7) are limited to vertical pressure vessels with small to medium diameters (1-10 ft) and moderate height-to-diameter ratio (H/d = 5-2). They are usually supported on structural steel or columns. If the operating temperature is high, thermal expansion of the shell cannot be absorbed by the supporting structure, and high stresses would consequently be induced in the shell, special sliding bearing plates are provided to reduce friction (see Fig. A1, Appendix A4). On small drums the top bar in Fig. 5.7 can be omitted. On larger, heavy vessels high localized stresses would develop at the tips of the gusset

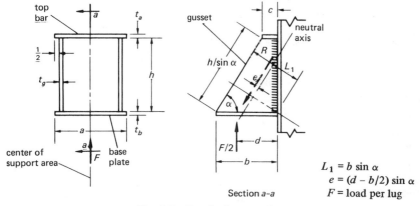

Fig. 5.7.   Detail of support lug.

plates. *The base plate* has to accommodate one or several anchor bolts. It could be stress checked as a uniformly loaded rectangular plate with one edge free and other three supported. *The gusset plates* can be assumed to be eccentrically loaded plates. Combined stress in one gusset due to the load $F/2$ consists of bending stress and direct compressive stress. From Fig. 5.7,

$$(F/2) d = Rd \sin \alpha \quad \text{and} \quad R = F/2 \sin \alpha$$

and the maximum compressive stress is

$$S_c = (R/L_1 t_g) + (6Re/L_1^2 t_g).$$

The gusset thickness t_g is then given by

$$t_g = F(3d - b)/S_a b^2 \sin^2 \alpha$$

where S_a is the allowable stress in compression.

The top bar can be assumed to be a simple supported beam with uniformly distributed load Fd/h. The bar thickness t_a is then given by

$$S = 6M/t_a c^2 = (6/t_a c^2)(Fda/8h)$$

or

$$t_a = 0.75 (Fda)/S_b c^2 h$$

where

$c = 2$ in. min. and $8t_a$ max.
S_b = the allowable stress in bending for the bar material.

The weld attaching the lug to the vessel carries the vertical shear load F and moment Fd.

Example 5.1. Design the support lugs for the vertical drum shown in Fig. 5.8, with the following design data:

design pressure: $P = 420$ psig
design temperature: 150°F
shell plate allowable stress: $S_a = 15,000$ psi
structural steel allowable stress: $S_a = 20,000$ psi
weights: operating, $W_o = 122$ kips $> W_T$
          empty, $W_e = 47$ kips
wind design pressure: 40 psf, shape coefficient 0.6

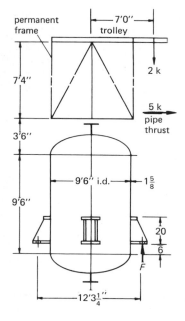

Fig. 5.8.

effective drum diameter for wind loads: D_e = 12.5 ft

wind pressure per foot of vessel height: w = 40 × 0.6 × 12.5

$$= 300 \text{ lb/ft}$$

4 lugs required.

1. *Wind loads.*

DESCRIPTION	SHEAR (kips)	ARM (ft)	M (kips-ft)
Wind on vessel			
300 lb × 10 ft	3.0	5	15.0
Wind on frame			
40 × 8 ft × 3 ft	1.0	15.5	15.5
Trolley 2^k × 7 ft	–	–	14.0
Pipe thrust	5.0	15.0	75.0
Total at lug	9.0	–	119.5

2. *Maximum F force on one lug.*

$$F = (4M/Nd_b) + (W_o/N)$$

$$= (4 \times 119.5/4 \times 12.3) + (122/4) = 9.7 + 30.5$$

$$= 41 \text{ kips.}$$

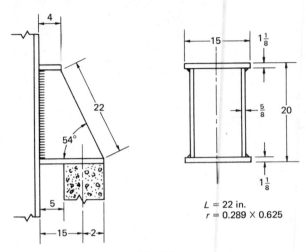

Fig. 5.9.

By inspection, uplift < 0. Use minimum bolt size per lug = 1 in. diameter, one per lug.

3. *Support lug (Fig. 5.9)*.
(a) Base plate: Bearing pressure,

$$p = 41,000/12 \times 15 = 228 \text{ psi}.$$

Maximum stress in base plate [46],

$$\sigma = \beta p b^2/t^2 = 0.72 \times 228 \times 12^2/1.125^2$$
$$= 18,700 \text{ psi} < 22,000 \text{ psi}.$$

For $15/12 = 1.25, \beta = 0.72$.
(b) Top bar plate thickness:

$$t_a = 0.75 \times 41,000 \times 15 \times 15/20,000 \times 4^2 \times 19$$
$$= 1.14 \text{ in}.$$

Use $1\frac{1}{8}$-in.-thick plate 4 in. wide
(c) Gusset plate thickness:

$$t_g = 41,000(3 \times 15 - 17)/9,885 \times 17^2 \times \sin^2 54° = 0.61 \text{ in}.$$

where allowable compressive stress is

$$S_a = \frac{18,000}{1 + \dfrac{1}{18,000}\left(\dfrac{L}{r}\right)^2} = 9,885 \text{ psi.}$$

(d) Size of the lug-to-shell weld:

shear: $f_1 = F/L_w = 41,000/(2 \times 18 + 2 \times 15) = 621$ lb/in.

bending: $f_2 = (41,000 \times 15)/[15 \times 19 + (19^2/3)] = 1,520$ lb/in.

combined: $f = (f_1^2 + f_2^2)^{1/2} = 1,640$ lb/in.

weld leg: $w = 1,640/0.55 \times (20,000 \times 0.6) = 0.249.$

Use $\frac{1}{4}$ in. fillet weld both sides for top bar and base plate, $\frac{1}{4}$ in. 50 percent intermittent for gusset plates.

4. *Stresses in Shell (see Chapter 7).*

$$c = \tfrac{1}{2}(15 \times 20)^{1/2} = 8.8, \qquad \beta = c/r = 8.8/57.7 = 0.15,$$

$$\gamma = r/t = 57.7/1.625 = 35.5$$

Maximum bending stress:

$$\sigma_b = C_{Lt}\, \frac{M_L}{t^2 r \beta}$$

$$= 0.355(41,000 \times 15)/1.625^2 \times 57.7 \times 0.15$$

$$= 9,550 \text{ psi.}$$

Combined stress:

$$\sigma = 9,550 + (420 \times 57.8)/1.625 = 24,490 \text{ psi} < 2S_a$$

Supporting Lugs with Full Reinforcing Rings

If stresses in the shell at the supporting lugs are too high and a reinforcing pad cannot satisfactorily be used, or if even a small deflection of the shell is objectionable (as on internally insulated piping), full circular reinforcing rings are used.

As an additional safety factor the stiffening effect of the adjacent shell zone is neglected and the required force and stress analysis is simplified. There are several disadvantages to reinforcing rings as compared with reinforcing pads or increased shell wall thickness. Since rings are much stiffer than the shell they introduce discontinuity stresses when the shell is pressurized. At high operating temperatures they act as fins and have to be properly insulated. They protrude from the shell and can interfere with piping or other equipment. Material used is either low-carbon structural weldable steel or of the same quality as the shell. The following example will describe the design procedure for sizing the ring stiffeners.

Example 5.2. A vertical, 72-in.-o.d., internally insulated pipe section with two supporting lugs (Fig. 5.10) is to be supported on two lugs with maximum load $W = 10,000$ lb. Determine the size of the stiffener rings.

Description of the Procedure. The vertical force $W/2$ per lug can be assumed to be resisted by the shear in the welds connecting the gusset plates to the pipe. The moment $Wd/2$ is carried into the rings as the force P equal to $Wd/2(h + t_r)$ acting in the plane of the ring curvature. No loading in the plane perpendicular to the plane of the ring is assumed. (The ring design can be used as a circular girder for large-diameter storage tanks where they are designed to carry loads perpendicular to the plane of the ring curvature.) The force P acts radially outward on the lower ring and inward radially on the top ring, as shown in Fig. 5.11. If a ring is subjected to equally spaced forces P in its plane, the maximum bending moment M_1 is at the points of loads given by

$$M_1 = \tfrac{1}{2} PR[(1/\alpha) - \cot \alpha] = K_1 PR,$$

where R is the centroidal radius of the ring.

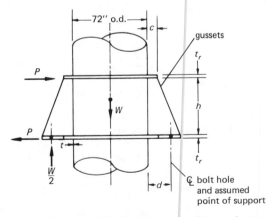

Fig. 5.10. Pipe section of Example 5.2. Top ring is continuous, bottom ring is continuous between the base plates.

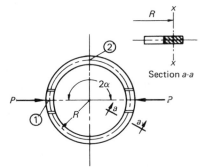

Fig. 5.11.   Top stiffening ring.

The angle between the loads P is equal to 2α. In addition to the bending moment M_1 a tangent (axial) thrust T_1 is induced in the ring at the load points:

$$T_1 = (P \cot \alpha)/2 = K_2 P.$$

T_1 stresses the ring bar in tension if P forces act outward or in compression if P forces act inward.

The resulting combined stress at the load point is

$$\sigma_1 = (T_1/a) + (M_1/Z_x) < S_a$$

where a is the cross-section area of the ring stiffener and Z_x is the ring section modulus about the axis x-x in Fig. 5.11. The bending moment at midpoints between the loads is

$$M_2 = (PR/2)[1/\sin \alpha - (1/\alpha)] = K_3 PR$$

the axial thrust is

$$T_2 = (P/2)(1/\sin \alpha) = K_4 P$$

and the combined maximum stress is

$$\sigma_2 = (T_2/a) + (M_2/Z_x) < S_a.$$

The coefficients K_1, K_2, K_3, and K_4 are evaluated in Table 5.3. The direct shear V at the load point is not algebraically additive to the bending and tension stresses and is usually small enough to be disregarded. The computed stress in the ring will be smaller than actual, since the shell effect is neglected; however, the forces P are transferred into the rings eccentrically, causing additional

Table 5.3. Values of K_1, K_2, K_3, and K_4.

NUMBER OF SUPPORTS	K_1	K_2	K_3	K_4
2	0.318	0.0000	0.182	0.500
3	0.189	0.289	0.100	0.577
4	0.137	0.500	0.070	0.707
6	0.089	0.866	0.045	1.000
8	0.066	1.207	0.034	1.306

stresses. Substituting the numerical values h = 12 in. and d = 6 in. at load point into the above equations: M_1 = 0.318(5,000 × 6/12) × 37.5 = 29,900 lb-in. and the required Z_x = 29,900/20,000 = 1.5 in.^{3}. Use 3-in.-wide × 1-in.-thick ring.

6

Design of Saddle Supports for Large Horizontal Cylindrical Pressure Vessels

6.1. GENERAL CONSIDERATIONS

Ideally, the saddle supports for long horizontal vessels should be located to cause the minimum stresses in shell and without required additional reinforcement. The actual location of the saddle supports is usually determined by the piping and platform layout. Most of the large horizontal cylindrical vessels are supported by two saddles, preferably with 120 degree contact angle, usually on concrete columns; sometimes the vessel can rest directly on two concrete piers. Any settlement of the supporting structure does not change the distribution of the load per saddle. (If multiple supports are used the reactions at saddles are computed from the theory of continuous beams and increased by 20–50 percent as a safety factor for settling of the supports.)

The cylindrical shell thickness is determined by the tangential stress due to the design pressure. Since the maximum longitudinal stress ($PR/2t$) is only half of the maximum tangential stress, one-half the shell thickness is available for the longitudinal bending stress due to weight or other loadings at the midspan or in the plane of the saddles, assuming the vessel to behave as its own carrying beam. The load must be transferred from the shell to the saddle. The saddle reactions are highly concentrated and induce high localized stresses in the shell. Their intensity changes with the distance of the saddles from the vessel end closures that reinforce the shell with their own stiffness, keeping the shell round. The exact analytical solution of the localized stresses in the shell above the saddles would be difficult and it is not at present available. The most frequently used approximate analysis, with theoretical conclusions supported by strain gauge measurements, was published in ref. 53 and is substantiated in ref. 102. The discussion which follows is a summary of this method.

6.2. MAXIMUM LONGITUDINAL BENDING STRESS IN THE SHELL

A horizontal vessel resting on two supports can be analyzed as a beam resisting the uniform load of the weight of the vessel and its contents by bending.

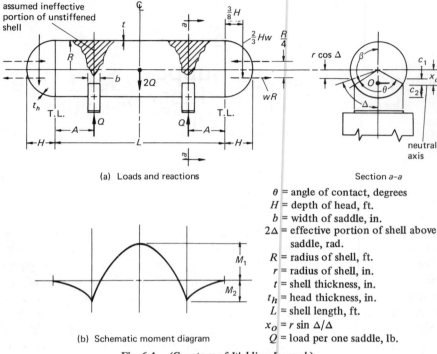

(a) Loads and reactions

Section *a–a*

θ = angle of contact, degrees
H = depth of head, ft.
b = width of saddle, in.
2Δ = effective portion of shell above saddle, rad.
R = radius of shell, ft.
r = radius of shell, in.
t = shell thickness, in.
t_h = head thickness, in.
L = shell length, ft.
x_o = $r \sin \Delta/\Delta$
Q = load per one saddle, lb.

(b) Schematic moment diagram

Fig. 6.1. (Courtesy of *Welding Journal*.)

If the total weight is $2Q$ and the equivalent length of the vessel is taken as $L + (4H/3)$ in Fig. 6.1, then the weight per linear foot of the beam is $w = 2Q/[L + (4H/3)]$. The liquid in the heads can cause only shear at the head–cylinder junction; this can be corrected for by adding a force couple acting on the head. The load and the moment diagrams are shown in Fig. 6.1. The maximum bending moments occur over supports and at the midspan.

Maximum Longitudinal Bending Stress S_1 in the Shell at the Midspan

The bending moment M_1 at midspan is

$$M_1 = w\left(\frac{L}{2} - A\right)\frac{1}{2}\left(\frac{L}{2} - A\right) - w\,\frac{2H}{3}\,A - wA\,\frac{A}{2} + wR\,\frac{R}{4} - \left(\frac{2wH}{3}\right)\left(\frac{3H}{8}\right)$$

$$= \frac{2Q}{L + 4H/3}\left[\frac{(L - 2A)^2}{8} - \frac{2}{3}HA - \frac{A^2}{2} + \frac{R^2 - H^2}{4}\right]$$

$$= \frac{QL}{4}\left\{\frac{1 + 2[(R^2 - H^2)/L^2]}{1 + (4H/3L)} - \frac{4A}{L}\right\}$$

$$= K_1(QL/4)\ \text{lb-ft.}$$

The value of K_1, the expression in braces, depends on A/L, with the distance A being the variable. All lengths are in feet and load Q in pounds. The section modulus is taken as $Z_1 = \pi r^2 t$ in.^{3}; the resulting bending stress S_1 at the midspan is given by

$$S_1 = \pm 12 M_1/Z_1 = \pm 12 QLK_1/4\pi r^2 t$$

or

$$S_1 = \pm 3QLK_1/\pi r^2 t \text{ psi.}$$

The maximum stress S_1 is either in tension (shell bottom) or compression (shell top).

Allowable Stress Limits. The tensile stress $+S_1$ combined with the longitudinal pressure stress $PR/2t$ should not exceed the allowable tensile stress of the shell material times the efficiency of the girth joints. The maximum compressive stress $-S_1$ occurs when the vessel is filled with the operating liquid and under atmospheric pressure. It should not exceed the Code maximum compressive stress for cylindrical shells.

Maximum Longitudinal Bending Stress S_1' in the Shell in the Plane of the Saddle

The bending moment in the plane of the saddle M_2 is equal:

$$M_2 = \frac{2Q}{L + 4H/3} \left[\frac{2H}{3} A + \frac{A^2}{2} - \frac{R^2 - H^2}{4} \right]$$

$$= QA \left\{ 1 - \frac{1 - (A/L) + [(R^2 - H^2)/2AL]}{1 + [4H/3L]} \right\} \text{ lb-ft.}$$

If the shell section above the saddle is *unstiffened* and forced to deflect, the high local tangential bending moments at the horn of the saddle render this section ineffective in bending to some degree. The effective arc of the unstiffened shell in bending is assumed to be

$$2\Delta = 2 \left[\frac{\pi}{180} \left(\frac{\theta}{2} + \frac{\beta}{6} \right) \right] \text{ rad}$$

and the effective section modulus Z_2 can be determined from Fig. 6.1:

$$Z_2 = I/c_1$$

$$= r^3 t \left\{ \frac{\Delta + \sin \Delta \cos \Delta - (2 \sin^2 \Delta/\Delta)}{r[(\sin \Delta/\Delta) - \cos \Delta]} \right\}$$

$$= \pi r^2 t \left[\frac{\Delta + \sin \Delta \cos \Delta - (2 \sin^2 \Delta/\Delta)}{\pi[(\sin \Delta/\Delta) - \cos \Delta]} \right] \text{ in.}^3$$

for the tension side. The stress S_1' is then

$$S_1' = 12M_2/Z_2$$

$$= \frac{3QL}{\pi r^2 t} \left\{ \frac{4A}{L} \left[1 - \frac{1 - (A/L) + (R^2 - H^2)/2AL}{1 + (4H/3L)} \right] \right.$$

$$\left. \times \left[\pi \frac{(\sin \Delta/\Delta) - \cos \Delta}{\Delta + \sin \Delta \cos \Delta - (2 \sin^2 \Delta/\Delta)} \right] \right\}$$

$$= +3QLK_1'/\pi r^2 t \text{ psi.}$$

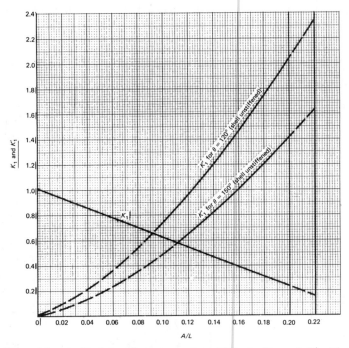

Fig. 6.2. Plot of longitudinal bending moment coefficients K_1 and K_1'. Dimension A should not exceed 20 percent of L. (Courtesy of *Welding Journal*.)

The value of K_1', the expression in the braces, varies mainly with the ratios A/L and H/L and the angle Δ. Both K_1 and K_1' can be plotted as constants for certain set of conditions. Assuming $H = 0$ for K_1 and $H = R$ for K_1' and $R/L = 0.09$ as a maximum value, the values of K_1 and K_1' are plotted in Fig. 6.2.

If the shell is *stiffened by a ring stiffener* in the plane of the saddle or by rings adjacent to the saddle, or if the saddle is close enough to the end closure ($A \leqslant R/2$) the effective angle 2Δ extends over the entire cross section, the section modulus is

$$Z = \pi r^2 t$$

and the maximum stress is

$$S_1' = \pm 12 M_2 / \pi r^2 t \text{ psi.}$$

Allowable Stress Limits. The tensile stress S_1' combined with the pressure stress $PR/2t$ should not exceed the allowable tensile stress for the shell material multiplied by the joint efficiency of the girth seam. Maximum compressive stress S_1' should be less than the Code allowable stress in compression.

6.3. MAXIMUM SHEAR STRESSES IN THE PLANE OF THE SADDLE

The distribution and the magnitude of the shear stresses in the shell produced by the vessel weight in the plane of the saddle will depend a great deal on how the shell is reinforced.

Shell Stiffened by a Ring in the Plane of the Saddle Away from the Head ($A > R/2$)

If the shell is made rigid enough with a stiffener the whole cross section will effectively resist the load-induced shear stresses.

The load V in Fig. 6.3. is the total vertical load on the left side of the ring in section *a-a* and is equal to

$$V = Q - \{2Q/[L + (4H/3)]\}(A + H) \doteq Q(L - 2A - H)/(L + H).$$

The shear force across the section per unit length of arc (shear flow) q_0 varies directly with the central angle ϕ and is given by

$$q_0 = V \sin \phi / \pi r \text{ lb/in.}$$

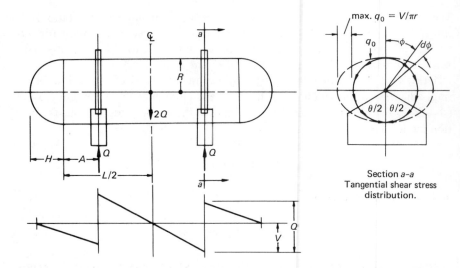

Section *a-a*
Tangential shear stress
distribution.

Fig. 6.3. Shear diagram for shell stiffened with a ring in the plane of the saddle. (Courtesy of *Welding Journal.*)

The total vertical load is

$$2 \int_0^\pi \left(\frac{V \sin \phi}{\pi r} \right) (\sin \phi)\, r\, d\phi = V.$$

The shear stress at any point adjacent to the stiffener will be

$$S_2 = q_0/t = V \sin \phi/\pi r t$$
$$= \frac{Q}{\pi r t} \left[\frac{L - 2A - H}{L + H} \right] \sin \phi = K_2 \frac{Q}{r t} \left[\frac{L - 2A - H}{L + H} \right].$$

The maximum value $K_2 = (\sin \phi)/\pi$ occurs at $\phi = 90$ degrees and $K_2 = 1/\pi = 0.319$, hence

$$S_2 = \frac{0.319 Q}{r t} \left[\frac{L - 2A - H}{L + H} \right] \text{ psi.}$$

Shell Not Stiffened by the Head ($A > R/2$) or Shell Reinforced by Two Ring Stiffeners Adjacent to the Saddle

Again the effective cross section of the shell resisting the shear stresses is taken to be reduced, with the maximum shear at the tip of the saddle. The arc of the

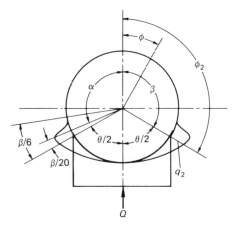

Fig. 6.4. Assumed distribution of shear stresses in an unstiffened shell above the saddle. (Courtesy of *Welding Journal*.)

effective cross section is assumed to be $2\Delta = (2\pi/180)[(\theta/2) + (\beta/20)]$ or $2(\pi - \alpha)$ in Fig. 6.4. The shear diagram is the same as for the stiffened shell, with the summation of the vertical shear on both sides of the saddle equal to the load Q:

$$2 \int_{\alpha}^{\pi} \left[\frac{Q \sin \phi_2}{r(\pi - \alpha + \sin \alpha \cos \alpha)} \right] r \sin \phi_2 \, d\phi_2 = Q.$$

With load V on one side of the saddle, the shear force is

$$q_2 = \frac{V \sin \phi_2}{r(\pi - \alpha + \sin \alpha \cos \alpha)}$$

$$= \frac{Q \sin \phi_2}{r(\pi - \alpha + \sin \alpha \cos \alpha)} \left[\frac{L - 2A - H}{L + H} \right]$$

and the shear stress is

$$S'_2 = (K'_2 Q/rt)[(L - 2A - H)/(L + H)] \text{ psi}$$

where $K'_2 = \sin \phi_2/(\pi - \alpha + \sin \alpha \cos \alpha)$. The maximum value of K'_2 occurs at $\phi_2 = \alpha$, for $\theta = 120$ degrees, $K'_2 = 1.171$ and for $\theta = 150$ degrees, $K'_2 = 0.799$.

The angles in formulas with trigonometric functions are in radians.

Shell Stiffened by Heads ($A \leqslant R/2$)

If the saddle is close to the end closure the shell is stiffened on the side of the head. A large part of the load Q inducing the tangential shears will be carried across the saddle to head and back to the head side of the saddle, with an assumed distribution as shown in Fig. 6.5.

The shear forces q_3 due to the saddle reaction Q are resisted by an arc $\pi - \alpha$ shell slightly larger than the contact angle $\theta/2$ of the saddle and are acting upward.

The shear forces in the reinforced section q_1 are acting downward. The resultants of the vertical components of both shear forces must be equal:

1. $\displaystyle 2 \int_0^\alpha \left(\frac{Q \sin \phi_1}{\pi r} \right) (\sin \phi_1) \, r \, d\phi_1 = Q(\alpha - \sin \alpha \cos \alpha)/\pi$

2. $\displaystyle 2 \int_\alpha^\pi \left(\frac{Q \sin \phi_2}{\pi r} \right) \left(\frac{\alpha - \sin \alpha \cos \alpha}{\pi - \alpha + \sin \alpha \cos \alpha} \right) (\sin \phi_2) \, r \, d\phi_2 = Q(\alpha - \sin \alpha \cos \alpha)/\pi.$

The shear stress in the shell arc from α to π is given by

$$S_2'' = q_3/t = \left(\frac{Q \sin \phi_2}{\pi r t} \right) \left(\frac{\alpha - \sin \alpha \cos \alpha}{\pi - \alpha + \sin \alpha \cos \alpha} \right)$$

or

$$S_2'' = \frac{K_2'' Q}{r t} \text{ psi}$$

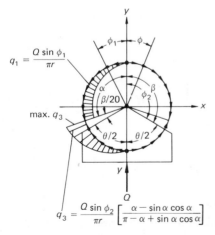

$$q_1 = \frac{Q \sin \phi_1}{\pi r}$$

max. q_3

$$q_3 = \frac{Q \sin \phi_2}{\pi r} \left[\frac{\alpha - \sin \alpha \cos \alpha}{\pi - \alpha + \sin \alpha \cos \alpha} \right]$$

ϕ = general angle

Fig. 6.5. Shear distribution in a shell reinforced by its heads. (Courtesy of *Welding Journal*.)

in the shell plate and

$$S_2'' = \frac{K_2'' Q}{r t_h} \text{ psi}$$

in the head plate, where

$$K_2'' = \left(\frac{\sin \phi_2}{\pi}\right)\left(\frac{\alpha - \sin \alpha \cos \alpha}{\pi - \alpha + \sin \alpha \cos \alpha}\right).$$

The value of K_2'' is maximum when $\phi_2 = \alpha$. For $\theta = 120$ degrees, maximum $K_2'' = 0.880$; for $\theta = 150$ degrees, maximum $K_2'' = 0.485$.

Allowable Stress Limits. The tangential shear stress should not exceed 0.8 times the allowable stress in tension.

6.4. CIRCUMFERENTIAL STRESS AT THE HORN OF THE SADDLE

The saddle reaction Q causes tangential shear forces in the shell cross section in the plane of the saddle, as evaluated in the previous section. These forces originate tangential bending moments and bending stresses in the shell, with the maximum bending stress at the tips of the saddle. Using the solution for bending moments in a ring with symmetrically applied tangential forces ($q = Q \sin \phi/\pi r$) an approximate solution can be derived for maximum stresses in an unstiffened shell and a shell reinforced by its end closure (see Fig. 6.6).

In the ring portion of uniform cross section fixed at the saddle horns, the circumferential moment M_ϕ at any angle ϕ is given by

$$M_\phi = \frac{Qr}{\pi}\left\{\cos\phi + \frac{\phi}{2}\sin\phi - \frac{3}{2}\frac{\sin\beta}{\beta} + \frac{\cos\beta}{2} - \frac{1}{4}\left(\cos\phi - \frac{\sin\beta}{\beta}\right)\right.$$

$$\times \left.\left[9 - \frac{4 - 6(\sin\beta/\beta)^2 + 2\cos^2\beta}{(\sin\beta/\beta)\cos\beta + 1 - 2(\sin\beta/\beta)^2}\right]\right\}.$$

The moment M_ϕ is maximum at $\phi = \beta$:

$$M_\beta = K_6 Q r \text{ lb-in.}$$

where K_6 is the expression in braces divided by π, with β substituted for ϕ.

The shear forces in an unstiffened shell are not distributed as in a stiffened shell, but are more concentrated at the tip of the saddle, giving bigger tangential shear stresses but smaller bending stresses.

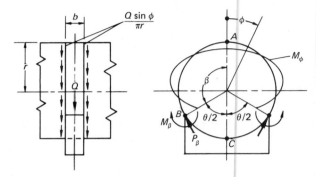

Fig. 6.6.   Load and circumferential moment M_ϕ diagram in the shell ring in the plane of the saddle. (Courtesy of *Welding Journal*.)

To be able to utilize the derived equation for M_ϕ and to bring the resulting bending stresses in the shell in agreement with the actual measured stresses, a fictitious resisting width of shell plate is taken as equal to $4r$ or $L/2$, whichever is smaller.

Further, to include the reinforcing effect of the head, the coefficient K_6 in the equation for M_β was adjusted to K_3:

$$M_\beta = K_3 Qr,$$

where

$$K_3 = K_6 \quad \text{for} \quad A/R > 1$$
$$K_3 = K_6/4 \quad \text{for} \quad A/R < 0.5$$

gradually increasing between the two values. A plot of K_3 is shown in Fig. 6.7.

The resulting bending stresses in psi are

$$S_b = \frac{M_\beta}{Z} = \pm \frac{6K_3 Qr}{4rt^2} = \pm \frac{3K_3 Q}{2t^2} \quad \text{for } L \geqslant 8R$$

$$S_b = \pm \frac{6K_3 QR}{Lt^2/2} = \pm \frac{12K_3 QR}{Lt^2} \quad \text{for } L < 8R.$$

The compressive stress due to the direct reaction P has to be added and was assumed to be equal to $Q/4$ for shells without stiffeners. However, the resisting width of the shell wall was taken only as the saddle width plus $5t$ on each side

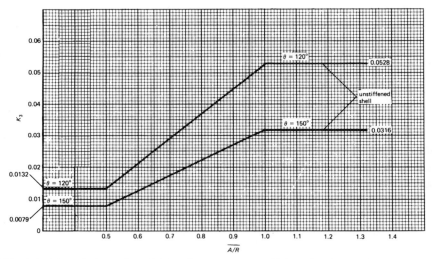

Fig. 6.7. Plot of circumferential bending moment coefficient K_3 for maximum stress at saddle horn. (Courtesy of *Welding Journal*.)

of the saddle, since the reaction P is located at the tip of the saddle. (A more liberal value of $0.78\sqrt{rt}$ each side of the saddle instead of $5t$ can be used [125].) The resulting direct stress is

$$S_c = \frac{-Q}{4t(b + 10t)} .$$

Since the sum of the two stresses is maximum when compressive, pressure stresses do not add to the above stresses. The resulting combined maximum stress at the saddle horn is given in psi by

$$S_3 = -\frac{Q}{4t(b + 10t)} - \frac{3K_3Q}{2t^2} \qquad \text{for } L \geqslant 8R$$

$$S_3 = -\frac{Q}{4t(b + 10t)} - \frac{12K_3QR}{Lt^2} \qquad \text{for } L < 8R.$$

Allowable Stress Limits. The computed maximum stress S_3 should not exceed 1.25 times the allowable stress for materials with equal tensile and compressive yield strength.

Note 1. Stress S_3 may be reduced by using a wear plate between the saddle flange and the vessel shell. The combined thickness of the wear plate and the shell can be used in the above formulas if the wear plate extends $0.10r$ in. above the saddle tips and the minimum width of the wear plate is $(b + 10t)$ in. The thickness of the added wear plate should not exceed the shell thickness t.

Note 2. Due to the high concentrated local stresses at the saddle horns, weld seams in the shell should be located away from them.

6.5. ADDITIONAL STRESSES IN A HEAD USED AS A STIFFENER

The shear forces q_1 and q_3 in Fig. 6.5 have variable horizontal components which cause additional secondary stress additive to the pressure stress. The induced stress in dished heads would be a combination of direct membrane and bending stresses and therefore difficult to evaluate analytically. However, if the head is replaced by a flat disk, the stresses become varying tension stresses across the entire vertical cross section, permitting an approximate solution applicable to practical design purposes.

The summation of the horizontal components across the section *y-y* in Fig. 6.5 is equal to the resultant force H:

$$H = \int_0^\alpha \left(\frac{Q}{\pi r} \sin \phi_1 \right) (\cos \phi_1) \, r \, d\phi_1$$

$$- \int_\alpha^\pi \left(\frac{Q}{\pi r} \sin \phi_2 \right) \left[\frac{\alpha - \sin \alpha \cos \alpha}{\pi - \alpha + \sin \alpha \cos \alpha} \right] (\cos \phi_2) \, r \, d\phi_2$$

$$= (Q/2)[\sin^2 \alpha/(\pi - \alpha + \sin \alpha \cos \alpha)] .$$

The total force is resisted by the area $2rt_h$ and the average tension stress will be equal to $H/2rt_h$. Assuming the maximum stress to be 1.5 times the above average stress, then

$$S_4 = (1.5/2rt_h)(Q/2)[\sin^2 \alpha/(\pi - \alpha + \sin \alpha \cos \alpha)]$$

or

$$S_4 = K_4 Q/rt_h$$

where

$$K_4 = \tfrac{3}{8} (\sin^2 \alpha)/(\pi - \alpha + \sin \alpha \cos \alpha).$$

For $\theta = 120$ degrees, $K_4 = 0.401$, and for $\theta = 150$ degrees, $K_4 = 0.297$.

Allowable Stress Limits.   The maximum stress S_4 combined with the pressure stress in the head should not be larger than 1.25 times the maximum allowable stress in tension for the head material.

6.6.   RING COMPRESSION IN THE SHELL OVER THE SADDLE

To compute the compression stresses in the shell band in contact with the saddle and corresponding saddle reactions a frictionless contact between the shell and the saddle flange is assumed.

The sum of the tangential forces acting at both edges of the saddle on the shell band directly over the saddle, as shown in Fig. 6.8, cause a ring compression stress in the shell band.

The maximum compression will be at the bottom at $\phi = \pi$. The saddle reactions R_ϕ are perpendicular to the shell passing through the center O and do not contribute to the shell compressive stress.

The total ring compressive shear force at any point A at an angle ϕ in the shell over the saddle will be equal to the summation of tangential shear forces on the shell arc above that point. The summation of the shear forces at both sides of the saddle is

$$T_\phi = - \int_\alpha^\phi q_3 r \, d\phi_2 - \int_\alpha^\phi q_1 r \, d\phi_1$$

$$= Q[(-\cos \phi + \cos \alpha)/(\pi - \alpha - \sin \alpha \cos \alpha)]$$

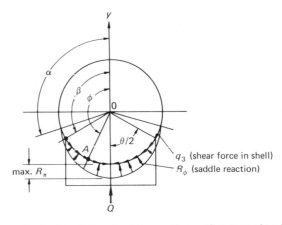

Fig. 6.8.   Compressive load in shell and saddle reactions.   (Courtesy of *Welding Journal.*)

with the maximum value at $\phi = \pi$:

$$T_\pi = Q(1 + \cos \alpha)/(\pi - \alpha + \sin \alpha \cos \alpha) \text{ lb.}$$

If the effective resisting shell width is taken as $b + 10t$, the maximum stress in compression is

$$S_5 = \frac{Q}{t(b + 10t)} \left[\frac{1 + \cos \alpha}{\pi - \alpha + \cos \alpha \sin \alpha} \right]$$

or

$$S_5 = K_5 Q/t(b + 10t) \text{ psi}$$

where for $\theta = 120$ degrees, $K_5 = 0.760$ and for $\theta = 150$ degrees, $K_5 = 0.673$.

Allowable Stress Limits. The maximum compressive stress S_5 should not exceed one-half the yield strength and is not additive to the pressure stress. If wear plate is used the combined thickness with the shell thickness can be used for computing stress S_5, provided the wear plate extends $r/10$ in. beyond the horn and its minimum width is $(b + 10t)$ in.

From the tangential shear forces T_ϕ the radial saddle reactions R_ϕ can be derived:

$$R_\phi r \, d\phi = T_\phi \, d\phi/2 + T_\phi \, d\phi/2$$
$$R_\phi = T_\phi/r = (Q/r)[(-\cos \phi + \cos \beta)/(\pi - \beta + \sin \beta \cos \beta)]$$

in lb/in., where the angle β was substituted for the angle α. For $\phi = \pi$ the reaction R_ϕ becomes maximum:

$$R_\pi = (Q/r)[(1 + \cos \beta)/(\pi - \beta + \sin \beta \cos \beta)].$$

The horizontal component of the reaction R_ϕ will cause tension across the saddle with the maximum splitting force F at the vertical centerline:

$$F = \int_\beta^\pi \frac{Q}{r} \left[\frac{-\cos \phi + \cos \beta}{\pi - \beta + \sin \beta \cos \beta} \right] (\sin \phi) r \, d\phi$$

$$= Q \left[\frac{1 + \cos \beta - (\sin^2 \beta)/2}{\pi - \beta + \sin \beta \cos \beta} \right]$$

or

$$F = K_8 Q$$

where for $\theta = 120$ degrees, $K_8 = 0.204$, and for $\theta = 150$ degrees, $K_8 = 0.260$. The effective resisting cross-sectional area of the steel saddle or of the reinforcing steel in a concrete support should be within $r/3$ distance from the shell.

Allowable Stress Limits.   The average tension stress due to the force F should not exceed two-thirds of the allowable tension stress of the saddle material, since the tangential bending stresses which have been neglected tend to increase the total splitting force.

6.7. DESIGN OF RING STIFFENERS

If the stresses due to the saddle reactions become excessive and wear plate dimensions too large, a ring stiffener in the saddle plane (preferred) or two ring stiffeners adjacent to the saddle are used.

Stiffener in the Plane of the Saddle

The distribution of the tangential shears transmitting the weight Q into the ring is known and shown in Fig. 6.9. The equation for the bending moment M_ϕ induced in the ring by shear forces is given in Section 6.4. The maximum moment occurs at $\phi = \beta$:

$$M_\beta = K_6 Q r$$

where for $\theta = 120$ degrees, $K_6 = 0.0528$, and for $\theta = 150$ degrees, $K_6 = 0.0316$.

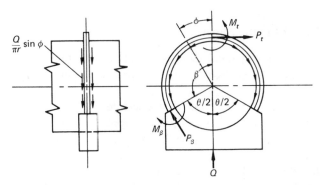

Fig. 6.9. Forces acting on stiffening ring in plane of saddle. (Courtesy of *Welding Journal*.)

The direct tangential force P_β at the point of $M_\phi = M_\beta$ is given by

$$P_\beta = \frac{Q}{\pi}\left[\frac{\beta \sin \beta}{2(1 - \cos \beta)} - \cos \beta\right] + \frac{\cos \beta}{r(1 - \cos \beta)}(M_\beta - M_t)$$

where M_t is the bending moment at $\phi = 0$ or

$$P_\beta = K_7 Q$$

where for $\theta = 120$ degrees, $K_7 = 0.340$, and for $\theta = 150$ degrees, $K_7 = 0.303$.

Assuming the size of the reinforcing ring of a cross-sectional area $a(\text{in.}^2)$ and of the section modulus $Z = I/c$ in.^{3} of the ring, the combined stress in the ring, maximum at $\phi = \beta$, is

$$S_6 = -(P_\beta/a) \pm [M_\beta/(I/c)]$$

or

$$S_6 = -(K_7 Q/a) \pm [K_6 Qr/(I/c)] \text{ psi.}$$

The added ring, usually of uniform cross-sectional area, is attached by welding to the outside of the shell so as not to obstruct the flow or prevent cleaning, and to the saddle horns.

Allowable Stress Limit. The stress S_6 in compression should not exceed one-half the yield strength in compression; the stress S_6 in tension, with added pressure tensile stress, should not exceed the allowable stress in tension.

Stiffeners Adjacent to the Saddle

The arrangement of two reinforcing rings adjacent to the saddle, for instance for a thin-wall cylindrical vessel supported directly on concrete piers, is seldom used.

If the stiffeners are placed close to the saddle (see Fig. 6.10) the rings will reinforce the shell section between. The tangential shear stresses will be distributed in rings as in Fig. 6.3, in section *a-a*, on the side away from the saddle. However, the shear distribution on the saddle side above the saddle will be similar to that shown in Fig. 6.4 for unstiffened shells.

The maximum bending moment in the rings occurs near the horizontal center-

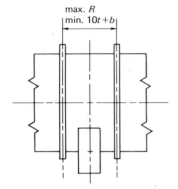

max. R
min. $10t + b$

Fig. 6.10.   Two stiffeners adjacent to the saddle.

line and is given by

$$M_\rho = K'_6 Qr/n$$

where for $\theta = 120$ degrees, $K'_6 = 0.0577$, for $\theta = 150$ degrees, $K'_6 = 0.0353$, and n is the number of rings used. The maximum tangential thrust P_ρ is given by

$$P_\rho = K'_7 Q/\pi n$$

where for $\theta = 120$ degrees, $K'_7 = 0.265$, for $\theta = 150$ degrees, $K'_7 = 0.228$, and n is the number of rings used. The maximum combined stress in the ring is

$$S'_6 = -(P_\rho/na) \pm [M_\rho/n(I/c)]$$

or

$$S'_6 = -(K'_7 Q/na) \pm [K'_6 Qr/n(I/c)] \text{ psi.}$$

The allowable stress limit is the same as for the ring in plane of the saddle. The K coefficients for the above formulas are summarized in Table 6.1.

6.8. DESIGN OF SADDLES

Design Remarks

The design loads for the saddle supports are the operating weight, combined with wind or earthquake loads, the friction force between saddles and foundation, and the test weight (see Fig. 6.11). The friction force is caused by expansion or contraction of the vessel shell if the operating temperature varies from

Table 6.1. Values of K-Coefficients in Formulas.

SUPPORT CONDITION	SADDLE ANGLE θ	LONGITUDINAL BENDING STRESS K_1 OR K_1'	TANGENTIAL SHEAR K_2, K_2', OR K_2''	CIRCUMFERENTIAL STRESS AT TOP OF SADDLE K_3 (2)	ADDITIONAL STRESS IN HEAD K_4	COMPRESSION IN SHELL OVER SADDLE K_5	RING STIFFENERS		
							CIRCUMFERENTIAL BENDING K_6 OR K_6'	DIRECT COMPRESSIVE STRESS K_7 OR K_7'	TENSION ACROSS SADDLE K_8
Shell unstiffened $A > R/2$	120°	(1)	1.171	0.0528	–	0.760	–	–	0.204
	150°	(1)	0.799	0.0316	–	0.673	–	–	0.260
Shell stiffened by heads $A \leqslant R/2$	120°	(3)	0.880	0.0132	0.401	0.760	–	–	0.204
	150°	(3)	0.485	0.0079	0.297	0.673	–	–	0.260
Shell stiffened by ring in plane of saddle	120°	(3)	0.319	–	–	–	0.0528	0.340	0.204
	150°	(3)	0.319	–	–	–	0.0316	0.303	0.260
Shell stiffened by rings adjacent to saddle	120°	(3)	1.171	–	–	0.760	0.0577	0.265	0.204
	150°	(3)	0.799	–	–	0.673	0.0353	0.228	0.260

1. See Fig. 6.2 for values of K_1 and K_1' or use the equations for K_1 and K_1'.
2. See Fig. 6.7 for values of K_3 at ratios A/R between 0.5 and 1.
3. See Fig. 6.2 for values of K_1 or use the equation for K_1.
(Courtesy of *Welding Journal*.)

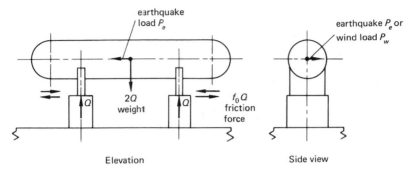

earthquake load P_a

earthquake P_e or wind load P_w

Q $2Q$ weight Q f_0Q friction force

Elevation   Side view

Fig. 6.11. Design loads for horizontal vessel support saddles.

atmospheric temperature. For small, temperature-induced shell-length changes saddles with the supporting structure can be made strong and flexible enough to resist the thermal forces. However, if a larger temperature movement of the vessel is expected, a special self-lubricating bearing plate with a low friction coefficient f_0 must be provided to reduce the expansion forces. A typical arrangement is shown in Fig. A2, Appendix A4.

Usually, unless the location specially requires it, the wind load can be disregarded in computations. If the stresses due to wind or earthquake loads are added to the other load stresses the allowable stresses can be increased by one-third.

Material. Steel plate materials commonly used in construction of saddles are A 283 grade A or C, or for low design temperature (below 32°F) A 516. Allowable design stresses per AISC Specifications:

tension: $S_a = 0.6S_y$,

shear: $S_s = 0.4S_y$,

bending: $S_b = 0.66S_y$.

Stress analysis. Saddle support components: the top flange, the web, the stiffeners, and the base plate with connecting welds are investigated for the maximum stresses.

The wear plate between the top flange and the vessel, if added to reduce the local concentrated stresses in the shell wall, is considered to be a part of the vessel.

Top flange. The design width of the top flange varies from 8 to 18 in. For computation of the thickness t_f, the flange can be considered as one-inch cantilever strip of length $b/2$ or more accurately c. However, with stiffeners and web

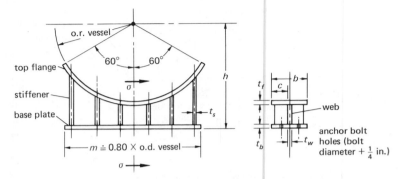

Fig. 6.12.   Typical detail of saddle support for large horizontal vessels.

welded to the flange, it can also be assumed as consisting of plates with three sides fixed and one free. The latter approach is more refined and gives a smaller thickness t_f, see ref. 52 and 109.

The web carries the compressive load from the weight as well as any bending due to thermal expansion force or earthquake.

Stiffeners reinforce the flange, the web, and the base plate. Together with the web they can be taken to be short columns, carrying the vertical load due to weight and bending moment due to the thermal expansion force at the saddle base. The approximate number of stiffeners is $N = (m/24) + 1$ (see Fig. 6.12), including the out-stiffeners. The stiffener thickness is $\frac{3}{8}$ in. minimum for vessels up to 6 ft in diameter and $\frac{1}{2}$ in. minimum for vessels with diameter larger than 6 ft.

Base Plate.   Compressive bearing pressure between the base plate and concrete is assumed to be uniform and limited by the allowable bearing pressure on the concrete foundation. The thickness t_b is computed the same way as the thickness for the flanges.

Anchor bolts.   Since the wind moment is not large enough to cause an uplift, the minimum size of the anchor bolts $\frac{3}{4}$–1 in. can be used to locate the saddles. However, they must be designed to overcome any thermal expansion or earthquake forces.

Welds.   Generally, the flange-to-shell weld is continuous all around. Welds connecting web to flange, web to base plate can be intermittent. Welds connecting stiffeners to the flange and to the base plates are continuous. Usually, the minimum weld sizes ($\frac{1}{4}$–$\frac{7}{16}$ in.) based on plate thicknesses are satisfactory.

Example 6.1.   Determine (a) the maximum stresses in shell, and (b) the required plate thicknesses for the support saddle for the horizontal drum shown in Fig. 6.13.

The design data are as follows:

design pressure: $P = 200$ psig
design temperature: 300°F
X-ray: full
saddle width: $b = 12$ in.
corrosion allowance: $\frac{1}{8}$ in.
material: shell,   SA 516 gr. 60
         saddle, SA 283 gr. C
vessel weight: empty = 130,000 lb
         liquid = 470,000 lb
         total weight = 600,000 lb
load per saddle: $Q = 300,000$ lb
saddle angle of contact: $\theta = 120$ degrees

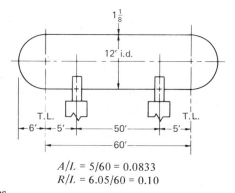

$A/L = 5/60 = 0.0833$
$R/L = 6.05/60 = 0.10$

Fig. 6.13.

(a) *Stresses in shell.*   (1) Maximum longitudinal stress due to bending at the midspan:

$$S_1 = \pm \frac{3K_1 QL}{\pi r^2 t} = \pm \frac{3(0.65)(300,000)(60)}{\pi (72.625)^2 (1)} = \pm 2,120 \text{ psi.}$$

K_1 taken from Fig. 6.2. Longitudinal stress in shell due to pressure:

$$SE = (P/2t)(r + 0.6t) = (200/2 \times 1)(72.125 + 0.6) = 7,273 \text{ psi.}$$

Total combined stress:

$$7,213 + 2,120 = 9,333 \text{ psi} < 15,000 \text{ psi.}$$

(2) Tangential shear stress in plane of saddle, $A > R/2, K_2' = 1.171$:

$$S_2' = \frac{K_2' Q}{rt} \left[\frac{L - 2A - H}{L + H} \right] = \frac{(1.171)(300,000)}{(72.625)(1)} \left[\frac{60 - 10 - 6}{60 + 6} \right]$$

$$= 3,220 \text{ psi} < 0.8 \times 15,000 = 12,000 \text{ psi.}$$

(3) Circumferential bending at horn of saddle $A/R = \frac{5}{6} = 0.833, K_3 = 0.041$. To reduce the stress use $\frac{1}{4}$-in. thick wear plate. $L = 60 > 8R = 48$.

$$S_3 = -\frac{Q}{4t(b + 10t)} - \frac{3K_3 Q}{2t^2} = -\frac{300,000}{(4)(1.25)(24.5)} - \frac{3(0.041)(300,000)}{2(1.25)^2}$$

$$= -2,450 - 11,808 = -14,258 < -1.25 S_a = 1.25 \times (-15,000) = -18,750 \text{ psi}.$$

(4) Ring compression in shell over the saddle, $K_5 = 0.760$:

$$S_5 = \frac{K_5 Q}{t(b + 10t)} = \frac{(0.760)(300,000)}{(1.25)(24.5)} = 7,445 \text{ psi} < S_y/2 = 16,000 \text{ psi}.$$

(5) Stresses in shell due to temperature expansion (Fig. 6.14):

Expansion: $\Delta l = \alpha l \Delta T = 7.10 \times 10^{-6} \times 50 \times 12 \times (300° - 70°) = 0.98 \text{ in}.$

Provide lubrite plates at one saddle as shown in Fig. A2, Appendix A4.

Friction coefficient: $f_0 = 0.10.$

Shear force at the saddle base: $f_0 \times Q = 30,000 \text{ lb}.$

$$x_0 = r \sin 60°/\text{rad}. \ 60° = 60 \text{ in}.$$

Bending moment $f_0 Q(84 - x_0) = 30,000 \times 24 = 720,000$ lb-in. is counteracted by the weight of the vessel, $Qb/2$. Use 8 $1\frac{1}{8}$-in.-diameter anchor bolts.

At the sliding saddle the nuts on bolts are hand tight and secured by tack welding, while the nuts and bolts at the fixed-end saddle are fully pretightened.

The friction force $f_0 Q$, in addition to the weight Q, is transmitted into the supporting structure.

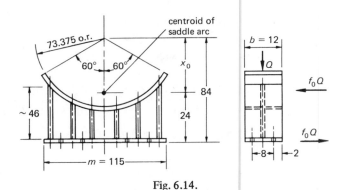

Fig. 6.14.

(b) *Saddle design.*   (6) Top flange thickness t_f.

$$P_\pi = \frac{Q}{r_0}\left[\frac{1 + \cos\beta}{\pi - \beta + \cos\beta\sin\beta}\right] = \frac{300,000}{73.5}\left[\frac{1 - \frac{1}{2}}{\pi - \frac{2}{3}\pi + (0.866)(-0.5)}\right]$$

$$= \frac{300,000}{73.5}\left(\frac{0.5}{0.615}\right) = 3,318 \text{ lb/in.}$$

Bending moment:

$$M_b = (P_\pi/b)(b/2)(b/4) = P_\pi b/8 = (3,318)(12)/8 = 4,980 \text{ lb-in.}$$

Flange thickness:

$$t_f = (6M/S_a)^{1/2} = [(6)(4,980)/(0.66)(30,000)]^{1/2} = 1.228 \text{ in.}$$

Use 1.25-in.-thick plate.

(7) Web thickness. Use $\frac{1}{2}$-in.-thick plate minimum. Calculate the maximum allowable height h of a 1-in.-wide strip column $\frac{1}{2}$-in. thick under P_π. Area $a = 1 \times \frac{1}{2}$ in.^{2} min. radius of gyration $k = 0.289 t_w$. From

$$\frac{P}{a} = \frac{18,000}{1 + \dfrac{1}{18,000}\left(\dfrac{h}{k}\right)^2} ,$$

GORDON – RANKINE

We get

$$h = t_w[(1,500/P_\pi)(18,000t_w - P_\pi)]^{1/2}$$

$$= 0.5[(1,500/3,318)(18,000 \times 0.5 - 3,318)]^{1/2}$$

$$= 25 \text{ in.} > 8.25 \text{ in.}$$

(8) Base plate thickness. Maximum stress:

$$S = M/Z = \frac{1}{2}(Q/2m)(b/2)/(t_b^2/6)$$

or

$$t_b = [(Qb)/26,400m]^{1/2} = [(300,000)(12)/(26,400)(115)]^{1/2} = 1 \text{ in.}$$

Use 1-in.-thick plate. Bearing pressure:

$$300,000/(12)(115) = 220 \text{ psi} < 750 \text{ psi allowable.}$$

(9) Maximum horizontal splitting force:

$$F = K_8 Q = (0.204)(300,000) = 61,200 \text{ lb}.$$

Area required of bottom cross section:

$$F/(0.66 \times 18,000) = 5.5 \text{ in.}^2.$$

Area available (Fig. 6.15):

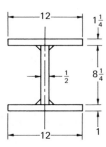

Fig. 6.15.

$$
\begin{aligned}
(12)(1) &= 12 \quad \text{in.}^2 \\
(0.5)(8.25) &= 4.12 \\
(12)(1.25) &= \underline{15.00} \\
& 31.12 \text{ in.}^2.
\end{aligned}
$$

(10) Welds connecting flange with wear plate.

$$\text{total length} = (73.5)(2)(\pi/3) + (2)(12) = 178 \text{ in.}$$

$$\text{shear in weld} = 30,000/178 = 170 \text{ lb/in.}$$

Use $\frac{1}{4}$-in. weld all around.
(11) Stiffeners.

$$N = m/24 + 1 = 6$$

including the end stiffeners. Spacing: about 23 in.

$$t_s = 0.5 \text{ in.}$$

Notes to saddle design computations. (1) A more accurate estimate of plate thicknesses and some resulting savings could be obtained by using more refined analytical procedures instead of simplifications made in the above; however, it is questionable whether additional work is justified in view of all assumptions that have to be made. (2) Structural steel is a comparatively cheap material. If the utmost use is made of the material, any later changes in the design are bound to be very costly or the safety factor will have to be reduced.

7
Local Stresses in Shells Due to Loads on Attachments

7.1. INTRODUCTION

In addition to the stress concentrations at the shell openings induced by the internal operating pressure, the localized stress effects of external loadings acting on nozzles and structural attachments such as supporting lugs are of primary importance.

High, concentrated stresses at such attachments due to combined internal pressure and local external loadings such as piping reactions on nozzles, can be a source of failures if proper reinforcement is not supplied. These stresses must, therefore, be evaluated. Because of the lack of geometrical symmetry a complete theoretical analysis of such localized stresses is too complicated for practical design work. A fully satisfactory analytical solution has not yet been accomplished at the present time.

However, within certain limits present theoretical analyses provide the designer with adequate design criteria. The methods described in sections 7.3, 7.4, and 7.6 can be used whenever external loads are transmitted into the shell, under internal pressure or no pressure, by a nozzle or a clip. The procedures are not applicable to shells with external design pressure, where shell buckling may occur.

7.2. REINFORCEMENT OF OPENINGS FOR OPERATING PRESSURE

Openings in vessel shells have to be reinforced for operating pressure (external or internal) in compliance with Code rules, which specify within certain limits the amount of removed metal by the opening to be replaced. It must be adequate for the design temperature and the design pressure.

If a tall column is to be hydrotested in a vertical position in the field, the reinforcement of all openings must be sufficient for the test pressure at the location. Essentially, the reinforcement provides sufficient strength for the weakened area to prevent excessive stress intensification around the opening.

Reinforcement is usually provided by a separate, welded reinforcing pad, by an increased amount of weld metal, by excess thickness in the shell and the

175

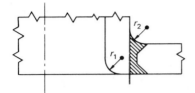

Fig. 7.1.

nozzle wall, or by using a heavier plate for the entire shell section or inserted locally around the opening.

Any shell thickness required for wind, weight, or earthquake loads in addition to the shell thickness needed for pressure cannot be used for reinforcement under operating conditions, or for any other sort of reinforcement.

Standard practice in vessel design is to replace all removed metal area by an opening:

$$A = d \times t$$

where

   A = the total cross-sectional area of reinforcement required
   d = the diameter of the finished opening (corroded)
   t = the nominal shell thickness, less corrosion allowance.

There are three important points to keep in mind for design of efficient reinforcing pads.

(1) Do not over-reinforce. Adding more material than required creates a too "hard" spot on the vessel, and large secondary stresses due to shell restraint can be produced in consequence.

(2) Place the reinforcing material adjacent to the opening for effectiveness. Code section UA-7 recommends placing two-thirds of the required reinforcement within a distance $d/4$ on each side of the opening.

(3) Use generous transition radii (r_1 and r_2 in Fig. 7.1) between the shell and the nozzle to minimize stress concentrations resulting from a grossly discontinuous junction under internal pressure. The results of tests show that even a minor outside radius r_2 has a large influence on the resulting peak stresses. The stress concentration factors at nozzles on a spherical shell subjected to internal pressure are plotted in ref. 57.

7.3. SPHERICAL SHELLS OR HEADS WITH ATTACHMENTS

A nozzle (or any other attachment) on a hemispherical head can be subjected to the following main external loads (see Fig. 7.2):

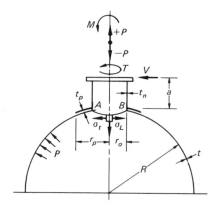

R = mean radius of the corroded spherical head
r_0 = outside nozzle radius
t_n = nozzle neck corroded thickness
t = head corroded thickness
t_p = reinforcing pad thickness; if used, it is additive to head thickness t in stress computations

Fig. 7.2.   Hemispherical head with a centrally located nozzle under principal loads.

1. radial load $\pm P$ (+ outward, - inward),
2. bending moment M,
3. torque T,
4. shear force V.

The shear stress in the shell at the nozzle o.d. due to the torque T is

$$\tau = (T/r_o)/2\pi r_o t = T/2\pi r_o^2 t$$

and the maximum shear stress in shell due to the force V is

$$\tau' = V/\pi r_o t.$$

These are usually small enough to be disregarded.

In most instances, the main governing stresses are associated with the bending moment M (including Va) and the radial force $\pm P$.

From Section 3.1 the biaxial principal stresses at the o.d. of the nozzle, σ_t and σ_L, can be expressed by the following general equations:

$$\sigma_t = \sigma_y = K_n \frac{N_y}{t} \pm K_b \frac{6M_y}{t^2}$$

$$\sigma_L = \sigma_x = K_n \frac{N_x}{t} \pm K_b \frac{6M_x}{t^2}$$

where

x = coordinate in longitudinal direction of shell.
y = coordinate in tangential (circumferential) direction of shell.
N_y = tangential stress resultant, lb/in.

M_y = tangential (circumferential) moment, lb-in./in.
N_x = longitudinal stress resultant, lb/in.
M_x = longitudinal moment, lb-in./in.
K_n, K_b = stress concentration factors depending on the material and geometry
of the junction. For static loads, K_n and K_b may be taken as equal
to one, since the peak stresses exceeding the yield point are redis-
tributed in ductile materials as used in vessel construction.

The stress resultants N_y and N_x and the bending moments M_y and M_x are
computed using thin-shell theory in ref. 54. To be of practical use, these theoretical results are expressed as a dimensionless general shell parameter plotted
against the attachment parameter. The graphs are updated in ref. 56, which is
currently used by designers for computing the stresses around attachments in
spherical and cylindrical shells. However, the designer is chiefly interested in
the maximum stresses in order to provide an adequate reinforcement. In order
to simplify computation the following steps can be taken to determine the
maximum stresses adjacent to the attachment.

1. *Maximum Stress Due to Radial Load $\pm P$.* The maximum stress in the shell
will be the longitudinal (meridional) stress σ_L in tension, given by

$$\text{max. } \sigma_L = \sigma_x = N_x/t + 6M_x/t^2$$

where for $+P$, N_x is tensile and for $-P$, N_x is compressive. This equation can be
restated as follows:

$$\sigma_L = \sigma_x = (P/t^2)[(N_x t/P) + (6M_x/P)]$$

where the values of $N_x t/P$ and M_x/P based on rigid-insert theory are given in
ref. 56. The initial equation can also be rewritten as

$$\sigma_L = C_p(P/t^2).$$

The values of $C_p = (N_x t/P) + (6M_x/P)$ are plotted in Fig. 7.3.

For $+P$ the maximum σ_L determined by the above formula can be added
directly to the membrane tensile stress due to internal pressure. However, for
$-P$ the stress resultant N_x is negative, causing compressive stress, and the maximum σ_L is

$$\sigma'_L = \sigma'_x = -(N_x/t) + (6M_x/t^2) = (P/t^2)[-(N_x t/P) + (6M_x/t^2)]$$

or

$$\sigma'_L = C'_p(P/t^2).$$

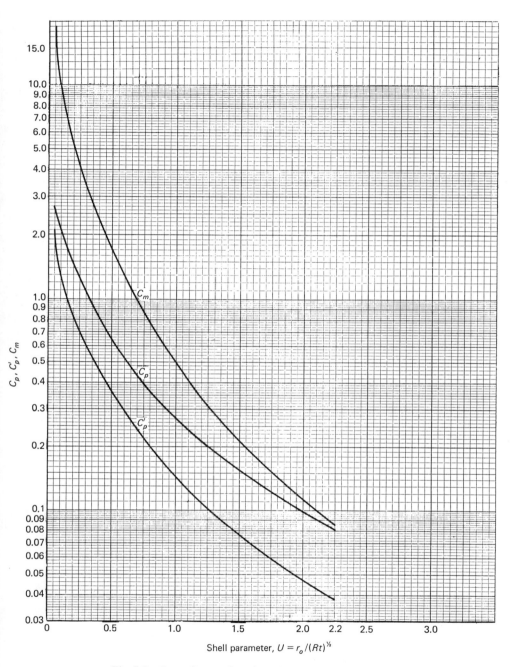

Fig. 7.3.   Stress factors C_p, C_p', and C_m for spherical shells.

The values of $C_p' = -(N_x t/P) + (6M_x/t^2)$ are also plotted in Fig. 7.3. The bending moment component M_x contributes considerably more to the resultant stress than does the N_x component. The resulting tensile stress σ_L' is directly additive to the pressure stress.

The stress factor $C_p'' = -(N_x t/P) - (6M_x/t^2) = -C_p$ would determine maximum compressive stress $-\sigma_L$, which is not additive to the pressure stress, but can govern under no-pressure conditions.

2. *Maximum Stress Due to Moment M.* The maximum stress due to M will again be the longitudinal stress σ_L. Based on the rigid-insert assumption, this stress is given by

$$\text{max. } \sigma_L = \sigma_x = (N_x/t) + (6M_x/t^2)$$
$$= (M/t^2 \sqrt{Rt})\,[(N_x t/M)\sqrt{Rt} + (6M_x/M)\sqrt{Rt}]$$

or

$$\sigma_L = C_m(M/t^2 \sqrt{Rt}) \quad \text{(in tension).}$$

The values of $C_m = [(N_x t/M)(Rt)^{1/2} + (6M_x/M)(Rt)^{1/2}]$ are plotted in Fig. 7.3. Here the bending tensile stress σ_L produced by M at point B in Fig. 7.2 is algebraically additive to the membrane stress due to the internal pressure.

The shell parameter (abscissa in Fig. 7.3) $U = r_0/(Rt)^{1/2}$ is for *round* attachments. The shell parameter for *square* attachments can be approximated by $U = c_1/0.875(Rt)^{1/2}$, where c_1 is equal to one-half the attachment side. For a *rectangular* attachment with maximum side ratio $c/b \leqslant 1.5$, c_1 can be roughly approximated by $c_1 = (ab)^{1/2}/2$.

The above procedure for finding the local stresses in spherical shells or hemispherical heads can be used in determining local stresses at attachments in crown sections of standard ellipsoidal or torispherical heads. The crown (dished) radius at the attachment is used as the mean radius R. All attachments should be placed away from the knuckle region, an area of concentrated pressure stresses.

Example 7.1. Determine if the reinforcement of the centrally located nozzle in a standard $2:1$ ellipsoidal head (Fig. 7.4) is satisfactory for the following design data:

Design pressure: $p = 250$ psig
Design temperature: $T = 450°F$
M and P are thermal expansion loads

Material: SA516-60, S_a = 15,000 psi
Reinforcing pad thickness: 0.625 in.

$$(t + t_p)/t_n = 1.25$$

Dished radius: $L = 0.9D_i = 65.4$ in.

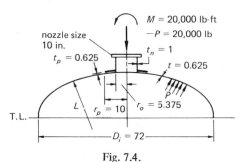

Fig. 7.4.

Minimum head thickness t required for pressure is

$$t = pD_i/(2SE - 0.2p) = (250 \times 72)/(2 \times 15{,}000 \times 1 - 0.2 \times 250) = 0.6 \text{ in.}$$

Use $\frac{5}{8}$-in.-thick plate.
 (a) *Checking stresses at nozzle outside diameter.*

$$U = r_o/[L(t + t_p)]^{1/2} = 5.375/(65.4 \times 1.25)^{1/2} = 0.6$$

Stress due to P:

$$\sigma'_L = C'_p P/(t + t_p)^2 = 0.29 \times 20{,}000/1.25^2 = +3{,}712 \text{ psi.}$$

Stress due to M:

$$\sigma''_L = C_m M/(t + t_p)^2 [R(t + t_p)]^{1/2}$$
$$= 1.21 \times 20{,}000 \times 12/1.25^2 \times 9.04 = +20{,}560 \text{ psi.}$$

Stress due to internal pressure p:

$$\sigma_L = pL/2(t + t_p) = +6{,}540 \text{ psi.}$$

Total stress:

$$\sigma = 30{,}812 \text{ psi} < 1.25(15{,}000 + 15{,}000) = 37{,}500 \text{ psi.}$$

 (b) *Checking the stresses at reinforcing-pad outside diameter.*

$$U = r_p/(Lt)^{1/2} = 10/(65.4 \times 0.625)^{1/2} = 1.56.$$

Stress due to P:

$$\sigma_L' = C_p'P/t^2 = 0.07 \times 20,000/0.625^2 = +3,584 \text{ psi}.$$

Stress due to M:

$$\sigma_L'' = C_pM/t^2(Lt)^{1/2}$$
$$= 0.19 \times 20,000 \times 12/0.625^2 \times (65.4 \times 0.625)^{1/2} = 18,260 \text{ psi}.$$

Stress due to internal pressure p:

$$\sigma_L = pL/2t = 250 \times 65.4/2 \times 0.625 = 13,000 \text{ psi}.$$

Total stress:

$$\sigma = 34,844 \text{ psi} < 1.25 \times (15,000 + 15,000) = 37,500 \text{ psi}.$$

7.4. CYLINDRICAL SHELL WITH ATTACHMENTS

A nozzle (or any other attachment) on a cylindrical shell can be subjected to the following external loadings, as shown in Fig. 7.5:

1. radial load $\pm P$ (+ outward, - inward),
2. longitudinal moment M_L in plane xy,
3. tangential (circumferential) moment M_t in plane xz,
4. torque T in plane parallel to yz,
5. shear V (tangential V_t, longitudinal V_L).

The shear stress in shell due to the twisting moment T is

$$\tau = (T/r_o)/2\pi r_o t = T/2\pi r_o^2 t$$

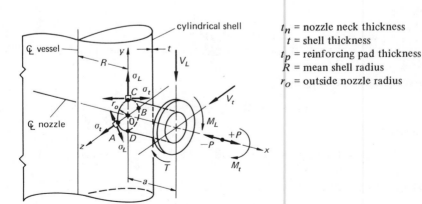

Fig. 7.5. Cylindrical shell with a radial nozzle under principal types of loads.

and the maximum shear in shell due to V_L or V_t is

$$\tau' = V_t/\pi r_o t \quad \text{or} \quad V_L/\pi r_o t.$$

τ and τ' are usually small enough to be disregarded.

The main stresses in most practical cases are associated with the moments M_L, M_t (including $V_L a$ and $V_t a$), and the radial load P. As in the case of spherical shells, the equations for σ_L and σ_t at the attachments were originally published in ref. 55 and are updated in ref. 56. The charts plotted in Figures 7.6, 7.7, 7.8, 7.9, and 7.10 allow direct determination of maximum stresses due to P, M_L, and M_t, important for design purposes. Since these charts were developed on the assumption that the attachment is rigid, the nozzle neck thickness has to be properly reinforced.

The parameters for cylindrical shells are as follows:

shell parameter: $\gamma = R/t$ or $R/(t + t_p)$ if reinforcing pad is used
square attachment: $\beta = c/R$ where c is the half-length of the loaded square area
cylindrical attachment: $\beta = 0.875 r_o/R$
rectangular attachment: β, rectangular area has to be converted into equivalent square loaded area. For small side ratios ($a/b \leqslant 1.5$) the equivalent $c = (ab)^{1/2}/2$ can be used; for larger ratios consult ref. 56.

1. *Stresses Due to Radial Load P.* The largest stress in the shell occurs generally in the tangential direction at the edge of the attachment on the circumferential centerline.

As in the case of spherical shells, here the tangential stress can be expressed by a general equation:

$$\sigma_t = \sigma_\phi = (K_n N_\phi/t) \pm (K_b 6 M_\phi/t^2)$$

where ϕ is the cylindrical coordinate in circumferential direction of shell, and taking K_n, K_b equal to unity,

$$\sigma_t = \sigma_\phi = (N_\phi/t) \pm (6M_\phi/t^2)$$

or

$$\sigma_t = \sigma_\phi = (P/t^2)\{[N_\phi/(P/R)\gamma] + (6M_\phi/P)\}.$$

The values of $N_\phi/(P/R)$ and M_ϕ/P are given in ref. 56. The value N_ϕ is positive (in tension) for $+P$ loads and the stress factor $C_p = [N_\phi/(P/R)\gamma] + (6M_\phi/P)$ is plotted in Fig. 7.6. The maximum σ_t in tension is then given by

$$\sigma_t = C_p(P/t^2).$$

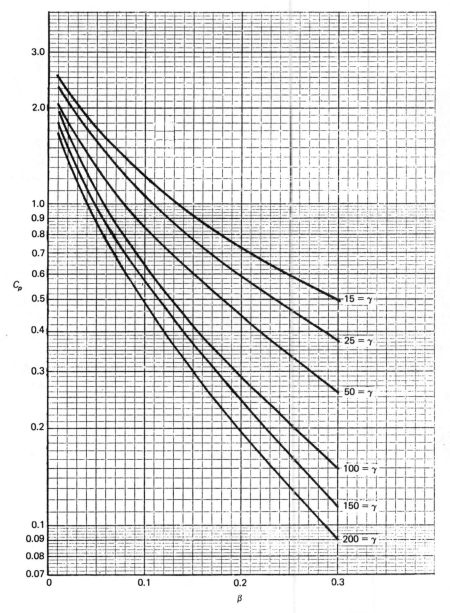

Fig. 7.6.   Stress factor C_p for cylindrical shells. TEN

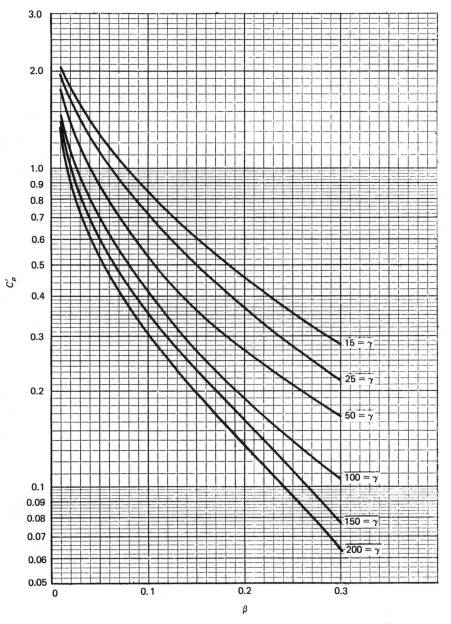

Fig. 7.7.   Stress factor C_p' for cylindrical shells.   *Comp.*

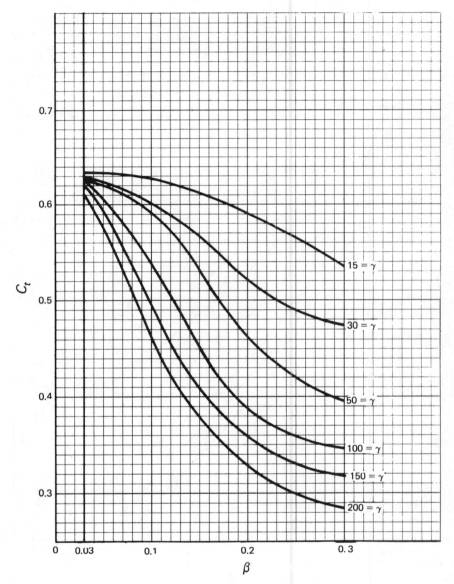

Fig. 7.8.   Stress factor C_t for cylindrical shells.

For $-P$ loads the value of N_ϕ is negative (in compression) and the stress factor $C_p' = [-N_\phi/(P/R)\gamma] + (6M_\phi/P)$ and plotted in Fig. 7.7. The maximum tensile stress is given by

$$\sigma_t = C_p'(P/t^2).$$

The σ_t obtained by the above equations can be algebraically added to the pressure stress.

2. *Stress Due to Tangential Moment M_t.* The maximum stress in the shell occurs generally in the tangential direction on the circumferential centerline of the attachment and can be written as

$$\sigma_t = \sigma_\phi = [N_\phi/(M_t/R^2\beta)] (M_t/R^2\beta t) + [M_\phi/(M_t/R\beta)] (6M_t/Rt^2\beta)$$
$$= (M_t/t^2 R\beta) \{[N_\phi/(M_t/R^2\beta)\gamma] + [6M_\phi/(M_t/R\beta)]\}$$

or

$$\sigma_t = C_t(M_t/t^2 R\beta)$$

in tension at point B. The values of $C_t = [N_\phi/(M_t/R^2\beta)\gamma] + [6M_\phi/(M_t/R\beta)]$ are plotted in Fig. 7.8 from data in ref 56. Stress σ_t is additive to pressure stress.

3. *Stresses due to Longitudinal Moment M_L.* The curvature of a cylindrical shell makes the longitudinal stiffness greater than the circumferential stiffness. The longitudinal stress can become larger than the tangential stress σ_t and govern the design condition of no internal pressure. However, the total combined tangential stress due to M_L and pressure stress will generally be larger than the combined longitudinal stress. The maximum combined stress occurs on the longitudinal centerline at the attachment–shell joint.

As before, the equation for the maximum tangential stress σ_t can be written as

$$\sigma_t = C_{Lt}(M_L/t^2 R\beta)$$

in tension at point C. The values of $C_{Lt} = [N_\phi/(M_L/R^2\beta)\gamma] + [6M_\phi/(M_L/R\beta)]$ are plotted in Fig. 7.9 using the data from the ref. 56.

The maximum longitudinal stress σ_L is given by

$$\sigma_L = C_{LL}(M_L/t^2 R\beta)$$

The values of $C_{LL} = [N_x/(M_L/R^2\beta)\gamma] + [6M_x/(M_L/R\beta)]$ are plotted in Fig. 7.10.

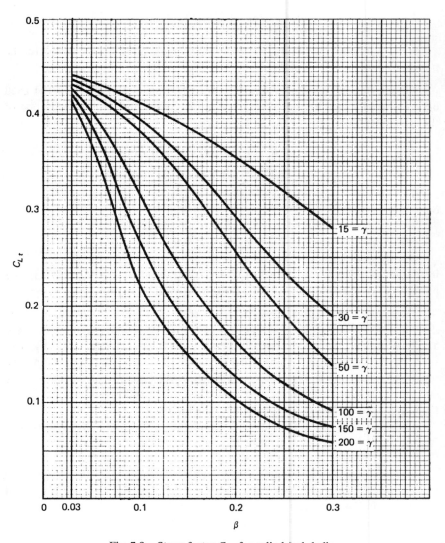

Fig. 7.9. Stress factor C_{Lt} for cylindrical shells.

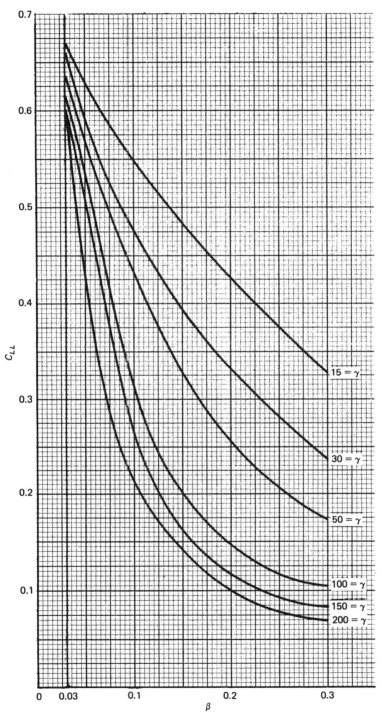

Fig. 7.10. Stress factor C_{LL} for cylindrical shells.

Example 7.2. A structural clip is welded to the cylindrical shell all around and subjected to external static loads, as shown in Fig. 7.11. Estimate the maximum stresses in shell.

$$S_a = 15,000 \text{ psi.}$$

Shell parameter:

$$\gamma = R/t = 60.25/0.5 = 120.5$$

Attachment parameters:

$$c = (14 \times 10)^{1/2}/2 = 5.92$$
$$\beta = 5.92/60.25 \doteq 0.10$$

Stress from $+P$:

$$\sigma_t' = C_p P/t^2$$
$$= 0.60 \times 1,000/0.5^2 = 2,400 \text{ psi}$$

Stress from M_L:

$$\sigma_t'' = C_{Lt}(M_L/t^2 R\beta)$$
$$= 0.3 \times 50,000/0.5^2 \times 60.25 \times 0.1 = 9,960 \text{ psi}$$

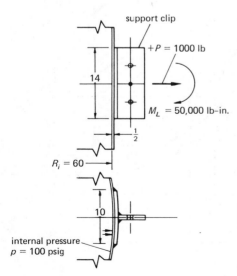

support clip

$+P = 1000$ lb

14

$M_L = 50,000$ lb-in.

$\frac{1}{2}$

$R_i = 60$

10

internal pressure
$p = 100$ psig

Fig. 7.11.

Stress from pressure:

$$\sigma_t = pR/t$$
$$= 100 \times 60.25/0.5 = 12,050 \text{ psi}$$

Combined stress:
$$\sigma = 2,400 + 9,960 + 12,050 = 24,410 \text{ psi} < 2S_a.$$

No increase in clip dimensions is required.

7.5 DESIGN CONSIDERATIONS

The following are some design considerations pertaining to sections 7.3 and 7.4.

1. *Rigid-insert theory* does not provide a good approximation of stresses in the shell if the nozzle body consists of a thin-wall, flexible pipe, even if the pipe is satisfactory for pressure and load stresses. The actual σ_L becomes smaller and the tangential stress σ_t becomes much larger. The maximum membrane stress resultants N_ϕ, N_x become much higher than those for a rigid insert. In addition, the peak stresses can occur across the base of the nozzle instead of in the shell. To prevent this, the section of the nozzle welded to the shell and under heavy external loads should be made thick enough so that the distribution of forces into the shell will approach the distribution assumed in rigid-insert theory, which assumes a continuous shell under the attachment. This is particularly important for larger-size nozzles.

2. *The computed stresses* are nominal and do not include the stress concentrations at the nozzle–shell junction. Peak stresses can become important for fatigue analysis. The stress concentration factors for cylindrical nozzles on spherical shells under pressure and various types of loads can be found in ref. 57.

3. *Size of reinforcing pad around attachment.* If the maximum stress at the attachment is too high, the shell must be reinforced by a reinforcing pad or the thickness of the reinforcing pad required for internal pressure must be increased. The width of the reinforcing pad is usually calculated so that the stresses at its edge will be below the allowable stress. This is done by assuming the reinforcing pad to be the attachment, as was done in Example 7.1. The width of the reinforcing pad, unless intermediate welds are used, should not exceed $16t_p$ or $16(t - \text{C.A.})$, otherwise the pad or shell carrying heavy bending stresses can fail in buckling. To avoid stress concentrations at corners of square or rectangular pads or structural clips under high loads (Fig. 7.11.) the corners are slightly rounded.

4. Generally, the loads are given as shown in Figs. 7.2 and 7.5. However, *if the loads are given in arbitrary planes*, for instance on a hillside nozzle, then

they must be decomposed vectorially into main vessel planes (meridional, tangential, and perpendicular to the meridional plane) before the stresses are computed. Each type of load has to be treated separately and the final stresses added. In most cases, the combined stress at the nozzle–shell junction will be the maximum. According to measurements, stress attenuates rapidly with distance from the junction.

5. *The maximum combined calculated stress* in the vessel wall cannot exceed the allowable stress for the design condition, as shown in Table 2.1.

6. *The minimum distance of the attachment from the edge* of the cylindrical shell (open or closed) in the above analysis is $R/2$. For smaller distances from the cylinder end the stresses decrease substantially, and can be adjusted by using plotted curves presented in ref. 55.

7.6. LINE LOADS

Structural plate clips welded to the shell transfer loads into the shell basically as line loads, usually along meridional lines or parallel circle lines. The analysis in the preceding sections deals with localized, uniformly distributed loadings over a finite, square area; the results of such an analysis cannot readily be applied to line loads. The analytical methods for solution of line-load stresses, as in ref. 59, are quite involved and prohibitively time consuming for practical design.

However, the stress at such clips under loads can be estimated by a simplified approach suggested in ref. 7. The maximum stress under the clips in Fig. 7.12, subjected to a radial load P, longitudinal moment M_L, and tangential moment M_t, is compared with the maximum stress under uniform circumferential line load, as shown in Fig. 7.13. The unit load f (lb/lin. in.) to be used for loads P, M_L, and M_t in the equations in Fig. 7.13 are determined from the following

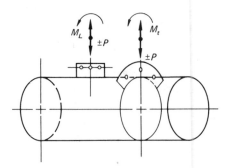

Fig. 7.12.

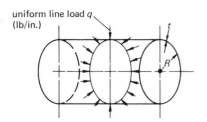

uniform line load q (lb/in.)

Longitudinal bending stress [46]:

$$\sigma_L = 3q/2\beta t^2 = 1.17q\,(Rt)^{1/2}/t^2$$

Membrane stress:

$$\sigma_{tm} = qR\beta/2t = 0.64q\,(Rt)^{1/2}/t^2$$

where $\beta = 1.285/(Rt)^{1/2}$

Fig. 7.13.

equations:

$$f_1 = M_L/Z_L$$
$$f_2 = M_t/Z_t$$
$$f_3 = P/L$$

where Z_L and Z_t are line section moduli in the longitudinal and tangential directions, respectively, and L is the length of the clip (see Table 10.3).

Substituting for q in the equation for σ_L in Fig. 7.13 and assuming that, because of the smaller rigidity of cylindrical shells in the tangential direction and uneven load distribution under the load P, the maximum induced stress under M_t and P will be 1.5-2 times larger than the stress computed by the formula, the following equations can be written for maximum stress:

$$S_1 = 1.17\,(Rt)^{1/2}f_1/t^2$$
$$S_2 = 1.75\,(Rt)^{1/2}f_2/t^2$$
$$S_3 = 1.75\,(Rt)^{1/2}f_3/t^2$$

The stresses due to M_L and P as well as the stresses from M_t and P, can be added algebraically. Membrane stresses due to internal pressure have to be superimposed on these stresses.

While this simplified procedure provides the designer with only comparative resulting stresses, they are within the range of stresses as measured under actual conditions. Furthermore, this simplified method affords the designer a quick check of problems that otherwise might remain unchecked and provides a safe design, as proven by experience.

Example 7.3. A structural clip is welded to a 10-ft-diameter cylindrical vessel shell with a design pressure 120 psig. The clip is subjected to the following

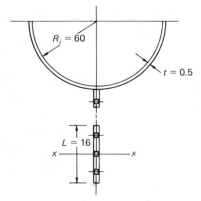

$$Z_x = L^2/6 = 42.6 \text{ in.}^2 \qquad\qquad \text{Fig. 7.14.}$$

mechanical loadings:

$$M_L = 2{,}000 \text{ ft-lb.}$$
$$P = +1{,}500 \text{ lb.}$$

Estimate the maximum stress in shell at clip.

Code allowable shell stress: $S_a = 15{,}000$ psi.
Maximum line loads (see Fig. 7.14):

$$f_1 = M_L/Z_x = 2{,}000 \times 12/42.6 = 565 \text{ lb/in.}$$
$$f_3 = P/L = 1{,}500/16 = 94 \text{ lb/in.}$$

Stress from P and M_L:

$$\begin{aligned}
\sigma_1 &= [1.17(Rt)^{1/2}/t^2](f_1 + 1.5f_3) \\
&= [(1.17)(60.3 \times 0.5)^{1/2}/0.5^2](710) \\
&= 18{,}245 \text{ psi.}
\end{aligned}$$

Stress from pressure:

$$\sigma_L = (120 \times 60.3)/(2 \times 0.5) = 7{,}240 \text{ psi.}$$

Combined maximum stress:

$$\sigma = 25{,}485 \text{ psi} < 2S_a.$$

8
Discontinuity Stresses

8.1. INTRODUCTION

As discussed in Chapter 3, industrial pressure vessels consist of axially symmetrical elements of different geometries, different shell thicknesses, or different materials. If the individual shell components are allowed to expand freely as separate sections under internal pressure, each such shell element would have an edge radial displacement ΔR and an edge rotation θ of the meridian tangent that would differ from the edge radial displacement and the edge rotation of the adjacent shell component. Since the shell elements form a continuous structure and must deflect and rotate together, at junctions these differences in radial displacements and rotations result in local shell deformations and stresses required to preserve the physical continuity of the shell. Stresses induced by such interaction of two shell components at their junction (an abrupt change in geometry of the vessel shell or a structural discontinuity) are called *discontinuity stresses*.

Discontinuity stresses can be analyzed by a general analytical method which is beyond the background and available time of the average designer. In practice, an engineering method that makes it possible to solve such complicated shell problems in a relatively short time is preferred. Since such method uses edge forces and edge moments as unknown quantities, it is called the *force method*. The force method offers a solution of local bending and shear stresses in boundary zones of junctions, where membrane theory is ineffective since it does not include any bending across the shell thickness or any perpendicular shear in its basic differential equations.

Discontinuity stresses themselves are usually not serious under static loads such as internal pressure with ductile materials if they are kept low by the design, but they become important under cyclic loads. Under a steady load they are considered to be of self-limiting nature, and higher allowable stress is permitted for the combined stress. Discontinuity stresses have to be superimposed on the membrane stresses caused by all other loadings, such as internal pressure, vessel weight, wind or earthquake loads, and thermal stresses. The general trend toward higher allowable stress values, larger vessels, and thinner shells of high-strength materials makes it more important to evaluate all local stresses, including discontinuity stresses.

The principles of the force method can best be explained using the following problem. A large-diameter pipe is rigidly anchored at intervals in concrete blocks, as shown in Fig. 8.1. If subjected to internal pressure of intensity P, the pipe will expand and the radial displacement will be ΔR. At section $O\text{-}O$, ΔR will be equal to zero since the pipe growth is prevented by the rigid concrete anchor and discontinuity stresses will be induced in the pipe wall. The force method divides such cases into two separate problems.

In the membrane solution [Fig. 8.1(b)] the built-in restraint at section $O\text{-}O$ is released (as in any statically indeterminate problem) and principal pressure

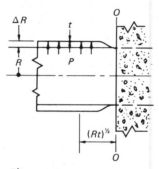

(a) $(Rt)^{1/2}$ = length of zone where discontinuity stresses are significant.

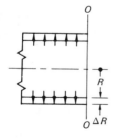

(b) Membrane solution

ΔR = radial displacement due to pressure P
θ = end rotation of shell due to pressure P is zero.

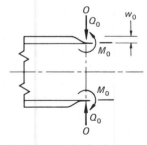

(c) Force-method solution

w_0 = radial displacement due to the end force Q_0 and the end moment M_0
θ_0 = end rotation due to Q_0 and M_0.

Fig. 8.1.

membrane stresses and radial deformation are computed:

$$\sigma_t = PR/t, \qquad \sigma_L = 0, \qquad \Delta R = R\sigma_t/E = R^2 P/tE.$$

In the force-method solution [Fig. 8.1(c)] the released edge of the pressurized shell at section O-O is subjected to the shear edge load Q_0 (lb/in.) and the edge moment M_0 (lb-in./in.) uniformly distributed along the pipe circumference at section O-O to bring the pipe edge in agreement with actual conditions. The unknowns to be computed here are the redundant force Q_0 and moment M_0. Once they are known the stresses in the pipe shell can easily be determined. The final state of stress at section O-O is the result of superposition of membrane pressure stresses plus the stresses due to Q_0 and M_0 or discontinuity stresses.

For most shell common forms the formulas for the edge deformations w_0 and θ_0 due to the edge loads Q_0 and M_0 have been derived using bending theory and are available to the designer. It is this availability of existing solutions that makes this method particularly attractive for a practicing engineer.

The influence coefficient is the deformation w_0 or θ_0 at the shell section edge due to the unit values of the edge loads ($Q_0 = 1$ lb/in.) and unit bending moment ($M_0 = 1$ lb/in.).

The maximum discontinuity stress (membrane plus bending) sometimes occurs at the loaded edge of the shell component where the total stress curve has a non-zero value slope. However, when solving for the maximum stress, in addition to the loaded edge stress the designer must always check the nearest location to the edge where the combined stress curve has a zero slope. Because of bending stresses both the inner and outer surfaces must be checked for maximum and minimum stresses. The location where the various combined stresses are maximum away from the loaded edge can be obtained by minimum–maximum theorem. The principal stresses developed in the shell at any location away from the loaded edge are expressed in terms of deformation $w(x)$ and bending moment $M(x)$, which in turn are expressed in terms of computed forces Q_0 and M_0 and the variable x, representing the distance from the loaded edge of the shell in the direction of decreasing stress. The location of the maximum stress can be obtained by equating the first stress derivative with respect to x to zero. However, analysis of this type is beyond the scope of this book. For all practical purposes the combined stress at the junction of a discontinuity can be taken as the maximum stress to be compared with the allowable stress.

In general, a shell component at a structural discontinuity can have the following boundary conditions:

1. free edge, $Q_0 = 0$ and $M_0 = 0$
2. fixed or built-in edge, $w_0 = 0$ and $\theta_0 = 0$, (see Fig. 8.1)
3. elastically built-in (elastically restricted) as at junctions of two shell components (see Fig. 8.3).

It should be pointed out that an important part of discontinuity stress analysis is the correct handling of signs. The usual sign convention is as follows:

(a) Radial shearing edge force Q_0 is positive when outward and tends to increase radius R.

(b) Longitudinal edge bending moment M_0 is positive when causing tension on the inside surface of the shell wall and tends to increase R. However, it may cause either plus or minus rotation θ_0.

(c) Radial deflection ΔR is positive when it increases R.

In solving the discontinuity problems calculator accuracy is required.

8.2. PROCEDURE FOR COMPUTING DISCONTINUITY STRESSES BY THE FORCE METHOD

The procedure for analyzing the discontinuity stresses by the force method can be summarized in the following steps:

1. A composite vessel shell under investigation is divided into simple shell elements, each with a single (or gradually changing) principal radius of curvature and thickness and made of a single material.

2. The edge radial displacement and rotations w_p and θ_p of the elements due to internal pressure are computed.

3. Any possible deformations due to other external mechanical or thermal loads are added: w_t and θ_t.

4. Edge shear force Q_0 and edge moment M_0 are applied. Directions are assumed at this point. Equilibrium of forces at the junction requires that shear Q_0 and moment M_0 at the edge of one component be equal and of opposite direction to those at the matching edge of the other component.

5. Using existing, tabulated formulas, the edge displacements w_0 and rotation θ_0 of each shell element at junctions due to Q_0 and M_0, respectively, are computed in terms of unknown edge forces Q_0 and M_0.

6. Total radial growth at one edge $w_1 = w_p + w_t + w_0$ is equated to the total radial growth w_2 at the second matching edge. Similarly, total edge rotation θ_1 of one edge is equated to total rotation of the second edge θ_2.

7. The two resulting linear equations are solved for the unknowns Q_0 and M_0. A negative sign in the resulting values of Q_0 or M_0 would indicate that the direction of Q_0 or M_0 is opposite to the direction originally assumed.

8. Discontinuity stresses caused by Q_0 and M_0 are computed.

9. To obtain the total stress at the junction, discontinuity stresses must be superimposed on pressure stresses, thermal stress, etc. Shear stresses perpendicular to the middle surface of the shell are neglected, since they are usually much

smaller than longitudinal and tangential stresses and a biaxial stress state is assumed.

8.3. CYLINDRICAL SHELLS

The following discussion gives the formulas for end deflections w_0 and end rotation θ_0 in the plane O-O in Fig. 8.2 under the radial shear line load Q_0 (lb/in.) and the edge moment M_0 (lb-in./in.), uniformly distributed on the circumference.

In most practical cases, the length of the cylinder exceeds the attenuation (decay) length $2(Rt)^{1/2}$, so the effects of Q_0 and M_0 at one end of the cylinder do not influence the effects of the loading conditions at the other end of the cylinder.

For thin-wall vessels the bending effects of the edge loadings, replacing the actual boundary or joint conditions, are confined to a narrow zone. In most rotationally symmetrical, geometrically complicated vessels this narrow zone can be replaced by an *equivalent cylinder* and an estimate of deformations as well as discontinuity stresses can be made on the basis of simple cylindrical formulas, as described below:

1. Radial displacement of the midsurface of a *closed-end* cylindrical shell due to the internal pressure P is given by

$$\Delta R = w = \frac{PR_i R}{Et} \, [1 - (\nu/2)] \text{ in.}$$

There is no end rotation due to the internal pressure: $\theta = 0$.

2. Radial displacement w_0 at section O-O due to the edge load Q_0 and bending moment M_0:

$$w_0 = \frac{Q_0}{2\beta^3 D} + \frac{M_0}{2\beta^2 D} \text{ in.}$$

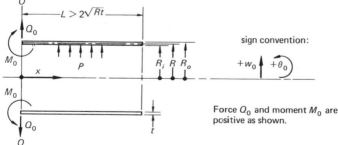

Fig. 8.2.

where

$$\beta = \left[\frac{3(1 - \nu^2)}{R^2 t^2}\right]^{1/4},$$

for $\nu = 3/10, \beta = 1.285/(Rt)^{1/2}$] and

$$D = Et^3/12(1 - \nu^2).$$

End rotation θ_0 due to Q_0 and M_0 at section O-O *clockwise* is given by

$$-\theta_0 = \frac{Q_0}{2\beta^2 D} + \frac{M_0}{\beta D} \text{ rad.}$$

3. The principal stresses due to edge loadings Q_0 and M_0 at section O-O are

$$\sigma_L = (N_L/t) \pm (6M_L/t^2) = \pm 6M_0/t^2 \quad \text{(bending only, } N_L = 0)$$
$$\sigma_t = (N_t/t) \pm (6M_t/t^2) = (Ew_0/R) \pm (6\nu M_0/t^2).$$

σ_t consists of the tangential membrane stress due to a change in radius R equal to w_0, which causes a tangential strain in shell equal to w_0/R per unit length of circumference and a stress Ew_0/R. σ_t can be either tensile or compressive, depending on the sign of deflection w_0. Further, since the sides of the unit element are here assumed to be restrained and unable to accommodate the rotation caused by the moment M_0, stresses equal to $\pm 6\nu M/t^2$ resulting from the Poisson effect are produced in the tangential direction.

4. Combined stresses due to internal pressure and edge loadings Q_0 and M_0 are

$$\sigma_L = (PR/2t) \pm (6M_0/t^2)$$
$$\sigma_t = (PR/t) + (Ew_0/R) \pm (6\nu M/t^2)$$
$$\sigma_r = -P/2 \text{ (average).}$$

5. Once edge loadings Q_0 and M_0 are known, deformations $w(x)$ and $\theta(x)$ and loadings $Q(x)$ and $M(x)$ at any distance x from the edge are given by

$$w(x) = \frac{e^{-\beta x} \cos \beta x}{2\beta^3 D} Q_0 + \frac{e^{-\beta x}(\cos \beta x - \sin \beta x)}{2\beta^2 D} M_0$$

$$-\theta(x) = \frac{e^{-\beta x}(\cos \beta x + \sin \beta x)}{2\beta^2 D} Q_0 + \frac{e^{-\beta x} \cos \beta x}{\beta D} M_0$$

$$M(x) = \frac{e^{-\beta x} \sin \beta x}{\beta} Q_0 + [e^{-\beta x}(\cos \beta x + \sin \beta x)]M_0$$

$$Q(x) = [e^{-\beta x}(\cos \beta x - \sin \beta x)]Q_0 - [2\beta e^{-\beta x} \sin \beta x]M_0.$$

Example 8.1. The top half of a non-Code vertical drum is constructed from 0.313-in.-thick stainless steel plate A240 Type 304L and the bottom half from 0.625-in.-thick (including corrosion allowance) carbon steel plate A285 gr. C. The following design data are given:

Drum i.d.: 120 in.
Maximum operating pressure: 70 psig
Periodic peak operating temperature: 650°F
$E_s = 25.1 \times 10^6$ and $E_c = 25.3 \times 10^6$ psi
$\alpha_s = 9.87 \times 10^{-6}$ and $\alpha_c = 7.33 \times 10^{-6}$ in./in./°F
Stainless steel type 304L:

$$S_a = 13,700 \text{ psi}$$
$$S_{atm} = 15,600 \text{ psi}$$
$$S_y = 25,000 \text{ psi}$$
$$S_y \text{ at } 650°F = 15,200 \text{ psi}$$

Determine the combined maximum stresses at the stainless-to-carbon steel shell junction in Fig. 8.3.

1. Shell deformations at junction due to the pressure P:

Radial deflection: $w_1 = PR_iR[1 - (\nu/2)]/E_ct_1 = +13,617 \times 10^{-6}$ in.
                       $w_2 = +27,330 \times 10^{-6}$ in.

Rotations: $\theta_1 = \theta_2 = 0$ rad.

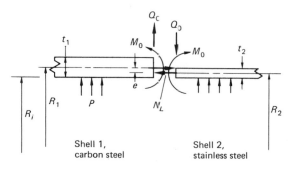

Assumed positive directions of Q_0 and M_0 as shown. Eccentricity e disregarded.

Shell 1, carbon steel

Shell 2, stainless steel

Fig. 8.3.

2. Deformations due to the rise in temperature $\Delta T = (650° - 70°) = 580°F$:

Radial growth: $\delta_1 = \alpha_1 R_1 \Delta T = +256{,}415 \times 10^{-6}$ in.
$\delta_2 = +344{,}335 \times 10^{-6}$ in.

Rotations: $\theta_1 = \theta_2 = 0$ rad.

3. Equating the summation of all deformations of the carbon steel shell to the summation of all deformations of the stainless steel shell end,

$$\beta_1 = 0.209, \quad D_1 = 0.568 \times 10^6$$
$$\beta_2 = 0.296, \quad D_2 = 0.071 \times 10^6.$$

Deformations: $w_1 + \delta_1 + \dfrac{Q_0}{2\beta_1^3 D_1} + \dfrac{M_0}{2\beta_1^2 D_1} = w_2 + \delta_2 - \dfrac{Q_0}{2\beta_2^3 D_2} + \dfrac{M_0}{2\beta_2^2 D_2}.$

Rotations: $\dfrac{Q_0}{2\beta_1^2 D_1} + \dfrac{M_0}{\beta_1 D_1} = \dfrac{Q_0}{2\beta_2^2 D_2} - \dfrac{M_0}{\beta_2 D_2}.$

and

$$Q_0 = 335 \text{ lb/in.}$$
$$M_0 = 360 \text{ lb-in./in.}$$

4. Total combined stresses σ_L and σ_t:
(a) Carbon steel shell:

$$\sigma_L = (PR_1/2t_1) \pm (6M_0/t_1^2) = 3{,}373 \pm 5{,}530$$
$$= +8{,}910 \text{ psi (inside)}$$
$$= -2{,}150 \text{ psi (outside)}$$
$$\sigma_t = (PR/t_1) + (E_c w_0/R_1) \pm (6\nu M_0/t_1^2)$$

where

$$w_0 = \frac{Q_0}{2\beta_1^3 D_1} + \frac{M_0}{2\beta_1^2 D} = +39{,}560 \times 10^{-6} \text{ in.}$$

Hence

$$\sigma_t = 6{,}756 + 16{,}600 \pm 1{,}660 = +25{,}020 \text{ psi (inside)}$$
$$= +21{,}700 \text{ psi (outside)}.$$

(b) Stainless steel shell:

$$\sigma_L = 6{,}726 \mp 22{,}050 = +28{,}780 \text{ psi (inside)}$$
$$= -15{,}324 \text{ psi (outside)}$$

$$\sigma_t = 13{,}450 - 25{,}715 \pm 6{,}615 = - 18{,}900 \text{ psi (outside)}$$
$$= -5{,}650 \text{ psi (inside)}$$

$$w_0 = - \frac{Q_0}{2\beta_1^3 D} + \frac{M_0}{2\beta_1^2 D} = -61{,}616 \times 10^{-6} \text{ in.}$$

Summary. The calculated σ_L in the stainless steel shell is far too high. The failure in the actual vessel occurred as a fissure parallel and close to the circumferential weld seam at a point where probably σ_L in combination with the residual weld stress first exceeded the ultimate fatigue strength. An occasional inspection would not prevent final breakthrough, since the fissure would originate on the inside of the vessel shell.

8.4. HEMISPHERICAL HEADS

The following discussion gives the formulas for the solution of effects of the uniformly distributed line edge shear Q_0 and the edge moment M_0 on a hemispherical head (Fig. 8.4). The sign convention used is as follows: w_0 is positive outward (increase in radius of curvature); θ_0 is positive outward.

1. Radial displacement of the midsurface due to the internal pressure P:

$$w = (PR^2/Et)\tfrac{1}{2}(1 - v) \text{ in.}$$
$$\theta = 0 \text{ rad.}$$

2. Radial displacement at section $O\text{-}O$ due to the loadings Q_0 and M_0 in Fig. 8.4:

$$w_0 = Q_0(2R\lambda/Et) + M_0(2\lambda^2/Et)$$

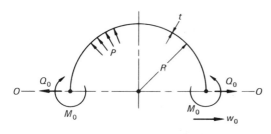

Edge loads Q_0 and M_0 are positive as shown.

Fig. 8.4.

where

$$\lambda = \beta R = [3(1 - \nu^2)/(Rt)^2]^{1/4} R.$$

End rotation θ_0 due to Q_0 and M_0:

$$\theta_0 = Q_0(2\lambda^2/Et) + M_0(4\lambda^3/REt).$$

3. The principal stresses due to edge loadings Q_0 and M_0 at section O-O are

$$\sigma_t = (N_t/t) \pm (6M_t/t^2) = (Ew_0/R) \pm (6\nu M_0/t^2)$$
$$\sigma_L = (N_L/t) \pm (6M_L/t^2) = \pm 6M_0/t^2.$$

4. Combined stresses due to internal pressure P and the edge loads Q_0 and M_0 at section O-O are

$$\sigma_t = (PR/2t) + (Ew_0/R) \pm (6\nu M_0/t^2)$$
$$\sigma_L = (PR/2t) \pm (6M_0/t^2)$$
$$\sigma_r = -P/2 \text{ (average)}.$$

The discontinuity stresses at the hemispherical head–cylindrical shell junction are less significant than at the semielliptical, torispherical, or conical head–cylindrical shell junctions. This could be an important factor in selecting a hemispherical head as an end closure for a large-diameter pressure vessel subject to high operating temperatures in addition to internal pressure.

Example 8.2. Determine the discontinuity stresses at junction of a top hemispherical head and cylindrical shell of a large-diameter vertical vessel (Fig. 8.5) with the following design data:

Design pressure: 150 psig,
Hydrotestpressure: 225 psig,
X-ray: full,
Design temperature: 700°F,
Inside radius: $R_i = 120$ in.,
Material: SA 515 gr. 60, $S_a = 14,300$ psi, $\nu = 0.3$, $E = 24.8 \times 10^6$ psi.

1. Shell thickness due to pressure is

$$t_s = \frac{PR_i}{SE - 0.6P} = \frac{150 \times 120}{14,300 - 0.6 \times 150} = 1.27 \text{ in.}$$

$$t_h = \frac{PR_i}{2SE - 0.2P} = \frac{150 \times 120}{2 \times 14,300 - 0.2 \times 150} = 0.63 \text{ in.}$$

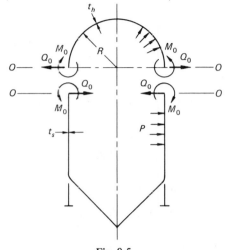

Fig. 8.5.

2. Edge deformations:

Shell—pressure:

$$w = (PR_i R/Et)[1 - (\nu/2)]$$

$$= \frac{0.85 \times 150 \times 120.63 \times 120}{24.8 \times 10^6 \times 1.27}$$

$$= 58{,}784 \times 10^{-6} \text{ in.}$$

$$\theta = 0 \text{ rad.}$$

Shell—edge loads:

$$w_0 = -(Q_0/2\beta^3 D) + (M_0/2\beta^2 D) = -96 \times 10^{-6} Q_0 + 10 \times 10^{-6} M_0$$
$$\theta_0 = +(Q_0/2\beta^2 D) - (M_0/\beta D) = 10 \times 10^{-6} Q_0 - 2 \times 10^{-6} M_0.$$

Head—pressure:

$$w = \tfrac{1}{2}(1 - \nu)(PR^2/Et) = 48{,}637 \times 10^{-6} \text{ in.}$$
$$\theta = 0 \text{ rad.}$$

Head—edge loads:

$$w_0 = +Q_0(2R\lambda/Et_h) + M_0(2\lambda^2/Et_h) = 273 \times 10^{-6} Q_0 + 40 \times 10^{-6} M_0$$
$$\theta_0 = +Q_0(2\lambda^2/Et_h) + M_0(4\lambda^3/REt_h) = 40 \times 10^{-6} Q_0 + 12 \times 10^{-6} M_0.$$

Total shell edge deflection = total head deflection:

$$58,784 - 96Q_0 + 10M_0 = 48,637 + 273Q_0 + 40M_0 .$$

Total shell edge rotation = total head edge rotation:

$$10Q_0 - 2M_0 = 40Q_0 + 12M_0.$$

Hence

$$M_0 = -108 \text{ lb-in./in.}$$

(in the opposite from that assumed), and

$$Q_0 = 36 \text{ lb/in.}$$

3. Combined stresses at section $O\text{-}O$:

Shell:

$$\sigma_L = (PR/2t_s) \mp (6M/t_s^2) = (150 \times 120.63)/(2 \times 1.27) \mp (6 \times 108/1.27^2)$$
$$= 7,146 \mp 404 = +6,742 \text{ psi (inside)}$$
$$= +7,550 \text{ psi (outside)}.$$

$$\sigma_t = (PR/t_s) + (Ew_0/R) \pm (6\nu M/t_s^2)$$
$$= 14,293 - 932 \mp 121 = +13,240 \text{ psi (inside)}$$
$$= +13,482 \text{ psi (outside)}$$

$$w_0 = -4,536 \times 10^{-6} \text{ in.}$$

Head:

$$\sigma_L = (PR/2t_h) \pm (6M_0/t_h^2) = 14,323 \mp 1,633 = +12,690 \text{ psi (inside)}$$
$$= +15,956 \text{ psi (outside)}$$

$$\sigma_t = (PR/2t_h) + (Ew_0/R) \pm (6\nu M/t_h^2)$$
$$= 14,323 + 1,131 \mp 490 = +14,995 \text{ psi (inside)}$$
$$= +15,944 \text{ psi (outside)}$$

$$w_0 = +5,500 \times 10^{-6}.$$

Check total deflection:

Shell: $58,784 - 4,536 = 54,248 \times 10^{-6}$ in.

Head: $48,637 + 5,500 = 54,137 \times 10^{-6}$ in.

Note: Any discontinuity stress will be further minimized by the plate taper at joint, as required by the Code welding detail in Fig. UW-13.1.

8.5. SEMIELLIPSOIDAL AND TORISPHERICAL HEADS

Semiellipsoidal Heads

In semiellipsoidal heads both principal radii of curvature are functions of the location (see Fig. 3.12.) The solution of edge deformations in terms of edge loadings becomes difficult, and bending theory does not provide ready-to-use, closed-form formulas for the deformations, as is the case with cylindrical shells and hemispherical heads. Approximation methods have to be used, which consume more time than is available to a vessel designer.

However, the most commonly used and commercially available 2:1 ellipsoidal head is satisfactorily designed using the Code formula with low acceptable discontinuity stresses.

If for some reason an ellipsoidal head with a higher R/h ratio is used and discontinuity analysis has to be performed, the following references will be of help to the designer: 59, 61, 63, 64, 101, 103, 107, 20, and 22.

Torispherical Heads

Discontinuity stresses are due to sharp changes in the radius of curvature at points a and 2 (Fig. 3.14). Since these points are in practice close together, the edge loadings affect each other to a large degree and again the analysis becomes prohibitively involved for routine vessel design work.

Since the total combined stresses in the knuckle region are several times higher than in standard 2:1 semiellipsoidal heads under the same pressure loading, the torispherical head is less suitable for high-pressure jobs (above 150 psi).

The references given in the discussion of semiellipsoidal heads can also be used as an assistance in estimating discontinuity stresses in torispherical heads.

8.6. CONICAL HEADS AND CONICAL REDUCERS
WITHOUT KNUCKLES

Conical heads and conical reducers without knuckles are frequently utilized in pressure vessel and piping design. The thickness of a conical head or a conical section under internal pressure P with a half apex angle $\alpha \leqslant 30$ degrees is computed by the simple Code membrane stress formulas and the Code rules requiring a reinforcement of the cone–cylinder junctions. No special analysis of discontinuity stresses is normally required.

The question now arises, how to check a cone–cylinder junction with α smaller than 30 degrees subject to external loads in addition to the internal pressure P, or with an angle α larger than 30 degrees. In such cases a more detailed analysis of discontinuity stresses at the cone–cylinder junction is in order.

The equations for the edge forces and moments for conical shells, being dependent on the cone apex angle (2α), are much more complicated than for cylindrical shells. While a number of exact solutions for the cone-shell edge displacements and forces under internal pressure and other loadings have been presented during the last decade, some in tabular form, they are suited rather to special jobs than to standard work in everyday pressure vessel and piping design.

The following discussion is an analysis of the localized stresses at the cone-cylinder junction, replacing the cone with an equivalent cylinder at the junction point under investigation. The resulting stress equations, based on much simpler cylindrical edge force formulas, are presented in forms readily applicable in standard design.

The stresses computed according to these formulas are in good, workable, overall agreement with the stresses calculated by more exact analytical methods. Also, design stress limits are suggested here to help the designer in evaluating a design for safety.

The whole analysis presented here is based on the ref. 60, carried out for different ratios of n = cone thickness/cylinder thickness.

Influence Coefficients

For the sake of convenience, the edge radial displacement ΔR and the edge rotation θ due to a unit edge load ($Q_0 = 1$ lb/in.) and a unit edge moment ($M_0 = 1$ lb-in./in.) are calculated for both cylinder and cone, replaced here by an equivalent cylinder (Fig. 8.6). These values are known as *influence coefficients*.

Cylinder. The radial deflection $w(=\Delta R)$ and rotation θ of the edge due to shear unit force $Q_0 = 1$ lb/in. is given by

$$w = 1/2D\beta^3$$

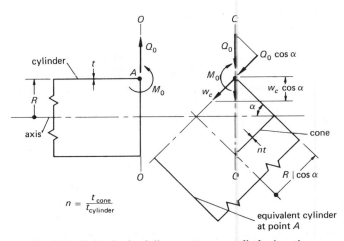

Fig. 8.6. Unit edge load diagram at cone-cylinder junction.

where $\beta = [3(1 - \nu^2)/R^2 t^2]^{1/4}$. For $\nu = 0.3$, $\beta = 1.285/(Rt)^{1/2}$.

$$\theta = 1/2D\beta^2 = w\beta, \qquad D = Et^3/12(1 - \nu^2).$$

The end deflection ΔR and rotation θ due to unit moment $M_0 = 1$ lb-in./in. are

$$\Delta R = 1/2D\beta^2 = w\beta \quad \text{and} \quad \theta = 1/\beta D = 2w\beta^2.$$

Cone (replaced by equivalent cylinder at point A). The radial displacement ΔR and rotation θ of the edge due to a unit shear force $Q_0 = 1$ lb/in. is given by

$$\Delta R = w_c \cos \alpha = [(1 \times \cos \alpha)/(2D_c\beta_c^3)] \cos \alpha = w(\cos^2 \alpha/n^3 k^3)$$

where

$$\beta_c = [3(1 - \nu^2)/(R/\cos \alpha)^2 (nt)^2]^{1/4} = \beta(\cos \alpha/n)^{1/2} = k\beta$$
$$D_c = n^3 D$$
$$k = (\cos \alpha/n)^{1/2}$$

$$\theta = (1 \times \cos \alpha)/(2D_c\beta_c^2) = w(\beta \cos \alpha/n^3 k^2).$$

The radial displacement ΔR and rotation θ of the edge due to a unit moment $M_0 = 1$ lb-in./in. are

$$\Delta R = (1/2D_c\beta_c^2) \cos \alpha = w(\beta \cos \alpha/n^3 k^2)$$
$$\theta = (1/D_c\beta_c) = w(2\beta^2/kn^3).$$

These results are tabulated in Table 8.1.

Table 8.1

	UNIT LOAD (1 lb/in.) PERPENDICULAR TO AXIS ON END OF		UNIT MOMENT (1 lb-in./in.) ON END OF	
	CYLINDER	CONE	CYLINDER	CONE
Unit shear	1	$1 \times \cos \alpha$	—	—
Unit moment	—	—	1	1
Radial displacement ΔR (in.) perpendicular to axis	w	$w \dfrac{\cos^2 \alpha}{n^3 k^3}$	$w\beta$	$w \dfrac{\beta \cos \alpha}{n^3 k^2}$
Rotation of meridian θ (rad.)	$w\beta$	$w \dfrac{2\beta^2 \cos \alpha}{n^3 k^2}$	$2w\beta^2$	$w \dfrac{2\beta^2}{kn^3}$
Circumferential membrane stress (psi)	$w\left(\dfrac{E}{R}\right)$	$w\left(\dfrac{E}{R}\right)\dfrac{\cos^2 \alpha}{n^3 k^3}$	$w\beta\left(\dfrac{E}{R}\right)$	$\dfrac{w\beta}{n^3 k^2}\left(\dfrac{E}{R}\right)\cos \alpha$

For $\nu = 0.3$, $w = (1/2D\beta^3) = (2.57/Et^3)(Rt)^{3/2}$.

Computation of Edge Forces f, F, and Edge Moment M

A closed-end conical reducer (Fig. 8.7) is separated from the adjacent cylindrical shells, by planes at junctures "L" and "S", and the acting end forces f, F and end moment M are applied as indicated.

Juncture L. From the equilibrium condition in the vertical direction:

$$f + F = (PR \tan \alpha)/2 = Z. \tag{1}$$

Using the formulas from Table 8.1, the total radial deflection of the cylindrical shell due to force f and moment M is equated to the total radial deflection of the cone due to the force F and the moment M. The radial displacements of both shells due to pressure are assumed approximately equal and cancel out.

$$fw - Mw\beta = [(Fw \cos^2 \alpha)/n^3 k^3] - [(Mw\beta \cos \alpha)/n^3 k^2] \tag{2}$$

where $\cos \alpha/n^3 k^2 = 1/n^2$. Similarly, the total rotation of the cylinder edge is equated to the total rotation of the cone edge. The end rotation of the conical shell due to pressure is neglected. This assumption is a source of discrepancy between approximate and more accurate analytical methods.

$$-fw\beta + Mw 2\beta^2 = [(Fw\beta \cos \alpha)/n^3 k^2] - (Mw 2\beta^2/n^3 k). \tag{3}$$

After simplifying, the result is a set of three linear simultaneous equations:

$$f + F = Z$$
$$f - M\beta = (Fk/n) - (M\beta/n)$$
$$-f + 2M\beta = (F/n^2) - (M 2\beta/kn^3)$$

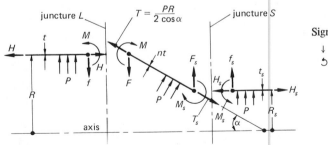

Sign convention used:
↓ + w deflection
↻ + θ rotation

R = mean radius of the larger cylinder
R_s = mean radius of the smaller cylinder

Fig. 8.7. Conical reducer separated into elements with applied edge forces and end moments positive as shown.

which can be solved for the three unknowns f, F, and M:

$$f = ZV_1$$
$$F = Z - f = Z(1 - V_1)$$
$$M = Z(2/\beta) V_2$$

where

$$V_1 = \frac{k[1 + n^2(1 + 2kn)]}{kn^4 + 2k^2 n^3 + 2kn^2 + 2n + k}$$

$$V_2 = \frac{kn[1 + V_1(n^2 - 1)]}{4(kn^3 + 1)} .$$

Both forces, f and F, are positive, acting in directions shown in Fig. 8.7 and causing compressive membrane stress in the shell. Moment M is also positive, producing bending stresses in compression in the outside of the shell.

Juncture S. The end force $T_s = (PR_s \tan \alpha)/2$ has the opposite direction to that at juncture L shown in Fig. 8.7, and so do forces f_s, F_s, and moment M_s.

Combined Stresses in Cylinder and Cone at Junctures L and S

Stresses in Cylinder at Juncture L. The combined longitudinal stress σ_L due to design pressure P and discontinuity moment M is given by

$$\sigma_L = (PR/2t) \mp (6M/t^2) = (PR/2t) \mp (PR/2)(6/t^2)(2\sqrt{Rt}/1.285) V_2 \tan \alpha$$

$$= \frac{PR}{t} (0.5 \mp 4.669 V_2 \sqrt{R/t} \tan \alpha).$$

The combined tangential (circumferential) membrane stress σ_{tm} (the Code average discontinuity hoop stress in par. UA5e) due to the pressure P, force f, and moment M, is given by

$$\sigma_{tm} = (PR/t) - fw(E/R) + Mw\beta(E/R)$$
$$= (PR/t) - ZV_1 w(E/R) + Z(2/\beta) w\beta(E/R) V_2$$
$$= (PR/t) - Zw(E/R)(V_1 - 2V_2)$$
$$= (PR/t) - (PR/2)(E/R)(6 \times 0.91/1.285^3)(R^{3/2} t^{3/2}/Et^3)(V_1 - 2V_2) \tan \alpha$$
$$= (PR/t)[1 - 1.285\sqrt{R/t}(V_1 - 2V_2) \tan \alpha].$$

In order to obtain the total combined stress σ_t, the bending stress ($\mp 6\nu M/t^2$) has to be added. However, since some movement of the differential shell element will reduce the bending stress in the tangential direction only the stress σ_{tm} is used in stress analysis.

The formulas for σ_L and σ_{tm} can be further simplified by rewriting the equations for σ_L and σ_{tm} as follows:

$$\sigma_L = (PR/t)(0.5 \mp X\sqrt{R/t}), \quad \text{where } X = 4.669\,V_2 \tan \alpha$$

$$\sigma_{tm} = (PR/t)(1 - Y\sqrt{R/t}), \quad \text{where } Y = 1.285(V_1 - 2V_2) \tan \alpha.$$

The coefficients X and Y, covering the majority of practical cases, are tabulated in Table 8.2 or the designer may plot them for different half apex angles α, as suggested in ref. 60.

Stresses in Cone Shell at Juncture L. The total combined longitudinal stress σ_L in the cone shell at juncture L is given by

$$\sigma_L = (PR/2tn \cos \alpha) \mp (6M/n^2 t^2) = (PR/2nt \cos \alpha) \mp 6Z(2/\beta)V_2/n^2 t^2$$
$$= (PR/2nt \cos \alpha) \mp (6/n^2 t^2)(PR/2) \tan \alpha(2/1.285)\sqrt{Rt}\,V_2$$
$$= (PR/t)[(0.5/n \cos \alpha) \mp (4.669/n^2)V_2\sqrt{R/t} \tan \alpha]$$
$$= (PR/t)[(0.5/n \cos \alpha) \mp U\sqrt{R/t}],$$

where

$$U = (4.669/n^2)V_2 \tan \alpha = X/n^2.$$

The membrane tangential stress σ_{tm} in cone shell at juncture "L" is:

$$\sigma_{tm} = (PR/nt \cos \alpha) - Fw(E/R)(\cos^2 \alpha/n^3 k^3) + Mw(\beta/n^3 k^2)(E/R) \cos \alpha$$
$$= (PR/nt \cos \alpha) - Z(1 - V_1)w(E/R)(k/n) + Z(2/\beta)V_2 w(\beta/n^2)(E/R)$$
$$= (PR/nt \cos \alpha) - (1.285PR/t)\sqrt{R/t} \tan \alpha \{[(k/n)(1 - V_1)] - (2V_2/n^2)\}.$$

Table 8.2.

$\alpha = 15°$						$\alpha = 30°$				
n	1	1.25	1.5	1.75	2	n	1	1.25	1.5	1.75	2
X	0.155	0.159	0.1573	0.151	0.142	X	0.325	0.335	0.330	0.317	0.300
Y	0.087	0.068	0.0647	0.058	0.054	Y	0.179	0.152	0.135	0.122	0.113
U	0.155	0.104	0.070	0.049	0.0356	U	0.325	0.214	0.147	0.104	0.074

$\alpha = 45°$						$\alpha = 60°$				
n	1	1.25	1.5	1.75	2	n	1	1.25	1.5	1.75	2
X	0.533	0.552	0.545	0.524	0.496	X	0.837	0.871	0.863	0.833	0.789
Y	0.293	0.255	0.218	0.197	0.180	Y	0.461	0.383	0.334	0.300	0.271
U	0.533	0.352	0.242	0.172	0.125	U	0.837	0.557	0.387	0.272	0.197

By a simple substitution for V_2, it can be shown that

$$\{[(k/n)(1 - V_1)] - (2V_2/n^2)\} = (V_1 - 2V_2)$$

so that

$$\sigma_{tm} = (PR/t)\{(1/n \cos \alpha) - [1.285(V_1 - 2V_2) \tan \alpha]\sqrt{R/t}\}$$

Table 8.3. Stresses at Cone-to-Cylinder Junctures.

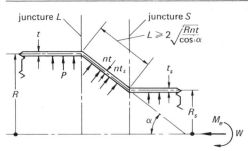

M_e = external moment at juncture
W = weight at juncture
$D = 2R$ or $2R_s$
t, t_s = corroded shell thickness

Positive (plus) sign means tension.
The upper sign specifies the stress on the outside surface.

$P_e = P + (4l/D)$, equivalent pressure for longitudinal stress.
$l = \pm(4M/\pi D^2) - (W/\pi D)$, unit longitudinal load, lb/in.

Juncture L
  Combined stress in cylinder.
    longitudinal:

$$\sigma_L = \frac{P_e R}{t}\left(0.5 \mp X\sqrt{\frac{R}{t}}\right)$$

    membrane tangential:

$$\sigma_{tm} = \frac{PR}{t}\left(1 - \frac{P_e}{P}Y\sqrt{\frac{R}{t}}\right)$$

  Combined stress in cone.
    longitudinal:

$$\sigma_L = \frac{P_e R}{t}\left(\frac{0.5}{n \cos \alpha} \mp U\sqrt{\frac{R}{t}}\right)$$

    membrane tangential:

$$\sigma_{tm} = \frac{PR}{t}\left(\frac{1}{n \cos \alpha} - \frac{P_e}{P}Y\sqrt{\frac{R}{t}}\right)$$

Juncture S
  Combined stress in cylinder.
    longitudinal:

$$\sigma_L = \frac{P_e R_s}{t_s}\left(0.5 \pm X\sqrt{\frac{R_s}{t_s}}\right)$$

    membrane tangential:

$$\sigma_{tm} = \frac{PR_s}{t_s}\left(1 + \frac{P_e}{P}Y\sqrt{\frac{R_s}{t_s}}\right)$$

  Combined stress in cone.
    longitudinal:

$$\sigma_L = \frac{P_e R_s}{t_s}\left(\frac{0.5}{n \cos \alpha} \pm U\sqrt{\frac{R_s}{t_s}}\right)$$

    membrane tangential:

$$\sigma_{tm} = \frac{PR_s}{t_s}\left(\frac{1}{n \cos \alpha} + \frac{P_e}{P}Y\sqrt{\frac{R_s}{t_s}}\right)$$

and finally

$$\sigma_{tm} = PR/t\,[(1/n\cos\alpha) - Y\sqrt{R/t}\,].$$

The coefficients U and Y for the majority of practical cases are tabulated in Table 8.2. To obtain the total tangential calculated stress σ_t, the bending stress $(\mp 6\nu M/n^2 t^2)$ has to be added.

Similarly, the Stresses at Juncture S Can Be Obtained. In conical reducers the maximum tangential compressive stress occurs at the large end, while the maximum tangential tension stress occurs at the small end under internal design pressure.

For the convenience of the designer all the above stress formulas are summarized in Table 8.3.

Example 8.3. In ref. 19, page 498, is an example of a rigorous analytical solution of discontinuity stresses at a cylinder-conical head junction with design data as shown in Fig. 8.8, with the following results

1. Computed stresses from ref. 19. (a) In the cylinder:

Total longitudinal stress:

$$\sigma_L = 5,690 \mp 36,560 = \begin{cases} -30,880 \text{ psi} \\ +42,260 \text{ psi}. \end{cases}$$

Membrane tangential:

$$\sigma_{tm} = 11,380 - 25.2(1,490) + 8.373(2,443)$$
$$= -5,712 \text{ psi}.$$

Total tangential:

$$\sigma_t = -5,712 \mp 9,140 = \begin{cases} -14,852 \text{ psi} \\ + 3,428 \text{ psi}. \end{cases}$$

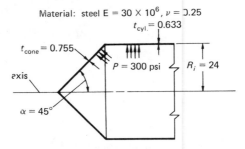

Material: steel $E = 30 \times 10^6$, $\nu = 0.25$

Fig. 8.8.   Design data for Example 8.3.

(b) In the cone:

Total longitudinal:

$$\sigma_L = 4{,}764 \mp 25{,}715 = \begin{cases} -20{,}951 \text{ psi} \\ +30{,}480 \text{ psi}. \end{cases}$$

Membrane tangential:

$$\sigma_{tm} = -5{,}898 \text{ psi}.$$

Total tangential:

$$\sigma_t = -5{,}989 \mp 5{,}902 = \begin{cases} -11{,}891 \text{ psi} \\ -\quad\;\; 87 \text{ psi}. \end{cases}$$

2. *Using the equivalent-cylinder approximation.* $R = 24.3$ in.; $n = 1.192$; $\nu = 0.3$. (a) In the cylinder:

Total longitudinal

$$\sigma_L = (PR/t)(0.5 \mp X\sqrt{R/t}) = (300 \times 24.3/0.633)(0.5 \mp 0.55\sqrt{24.3/0.633})$$

$$= 5{,}758 \mp 39{,}245 = \begin{cases} -33{,}487 \text{ psi outside} \\ +45{,}003 \text{ psi inside.} \end{cases}$$

Membrane tangential:

$$\sigma_{tm} = (PR/t)(1 - Y\sqrt{R/t}) = (300 \times 24.3/0.633)(1 - 0.256\sqrt{24.3/0.633})$$
$$= 11{,}516 - 18{,}267 = -6{,}750 \text{ psi}.$$

Total tangential:

$$\sigma_t = \sigma_{tm} \mp (6\nu M/t^2) = -6{,}750 \mp 0.3(39{,}245) = \begin{cases} -18{,}524 \text{ psi} \\ +\;\; 5{,}024 \text{ psi}. \end{cases}$$

(b) In the cone

Total longitudinal:

$$\sigma_L = (PR/t)[(0.5/n \cos \alpha) \mp U\sqrt{R/t}]$$
$$= (300 \times 24.3/0.633)[(0.5/1.192 \times 0.707) \mp 0.387\sqrt{24.3/0.633}]$$
$$= 6{,}832 \mp 27{,}614 = \begin{cases} -20{,}782 \text{ psi} \\ +34{,}446 \text{ psi}. \end{cases}$$

Membrane tangential:

$$\sigma_{tm} = (PR/t)[(1/n \cos \alpha) - Y\sqrt{R/t}\,]$$
$$= (300 \times 24.3/0.633)[(1/1.192 \times 0.707) - 0.256\sqrt{24.3/0.633}\,]$$
$$= 13,666 - 18,267 = -4,601 \text{ psi.}$$

Total tangential:

$$\sigma_t = -4,601 \mp 0.3 \times 27,614 = \begin{cases} -12,881 \text{ psi} \\ +\ \ 3,683 \text{ psi.} \end{cases}$$

The stresses as computed by this approximate analysis are in good and reasonable workable agreement with stresses calculated by the more exact method. Wherever the stresses differ they are on the conservative side. A better agreement would be reached by using $\nu = 0.25$ and $R = 24$ in.; however, since a Code designer would use $\nu = 0.30$ and midsurface radius in computations, the same values were used here for comparison.

Suggested Allowable Stresses and Stress Ranges at Cone–Cylinder Junctions

The calculated total longitudinal stress σ_L with predominant bending stress component can be very high and exceed the yield stress. As pointed out in ref. 60, the longitudinal stress cannot exist in the vessel wall in calculated intensities exceeding the yield point of the material.

A hydrostatic test can be assumed as the final fabrication operation here; it produces a local yielding in the vessel wall in the longitudinal direction, provided no local buckling occurs. (Of course, any visible localized deformations in the shell due to the hydrotest are objectionable, since they tend to increase the existing flaws in the base material and welds and initiate new cracks in the less ductile heat-affected zone adjacent to the weld as well as they introduce undesirable large residual stresses.)

When the shell is subsequently pressurized at the design pressure the induced σ_L should remain below the yield point. However, the relaxation of the longitudinal stress σ_L by yielding will increase the tangential component membrane stress by an amount equal to roughly one-half of the reduction in longitudinal stress [74].

Using this reasoning, the maximum allowable stresses per Code Division 1 par. UA53, $4S_a E$ in the longitudinal direction (membrane + discontinuity bending) and $1.5S_a E$ in the tangential direction (membrane hoop + discontinuity hoop), as compared with the computed localized stress σ_L and σ_{tm}, are assumed rather as the maximum allowable stress ranges.

The following are the maximum allowable design stress limits for σ_{tm} and σ_L based on experience and using Code Divisions 1 and 2 as a guide.

Maximum Allowable Tangential Membrane Stress σ_{tm} in Cone and Cylinder.
1. Design pressure plus any operating external mechanical load

$$\sigma_{tm} \leqslant 1.33 S_a \quad \text{or} \quad \sigma_{tm} + (\sigma_L/2) \leqslant 2 S_a < S_y.$$

2. Hydrotest pressure:

$$\sigma_{tm} \leqslant 2 S_{atm} \quad \text{or} \quad \sigma_{tm} + (\sigma_L/2) \leqslant 3 S_{atm} < S_u,$$

where S_a is the Code allowable stress at the design metal temperature and S_{atm} is the Code allowable stress at room temperature.

9
Thermal Stresses

9.1 GENERAL CONSIDERATIONS

Whenever there is a considerable difference between the vessel operating temperature and atmospheric (assembly) temperature, thermal expansion problems occur. Generally, the effects of externally applied mechanical loads and the effects of thermal expansion are separated and independently analyzed and only the final effects of total combined stresses are considered when comparing to maximum allowable stresses.

Thermal (or temperature) stresses in a structural member are caused by temperature changes and accompanying dimensional changes. In addition to being subject to a change in temperature, to develop thermal stresses the structural member must be restrained in some manner. The constraints of the member in the usual thermal stress problems that occur in the vessel design may be divided into external and internal constraints. This division is convenient, since the analytical procedures for solving the two types of problem differ. Internal constraints are usually the more difficult to solve.

In thermal analysis the first task is then to compute the elastic stresses in the part due to these constraints. It is assumed that the thermal stresses are within the elastic range on the stress–strain curve of the construction material. If the temperature is high enough the creep may become significant and have to be considered. Induced thermal stresses can easily exceed the yield strength of the material. However, thermal stresses are self limiting; this means that a small local plastic relaxation will reduce the acting force, as described in Chapter 2.

9.2. BASIC THERMAL STRESS EQUATIONS

A body can be visualized as composed of unit cubes of uniform average temperature. If the temperature of a unit cube is changed from T to T_1 and the growth of the cube is restrained, there are three cases to consider.

1. Restraint in the x-direction. A stress σ_x is induced:

$$\sigma_x = -\alpha E(T_1 - T)$$

in compression for $T_1 > T$, where α is the coefficient of thermal expansion (in./in./°F).

2. Restraint in x- and y-directions.

$$\sigma_y = \sigma_x = -\alpha E (T_1 - T)/(1 - \nu).$$

3. Restraint in all three directions (x, y, z).

$$\sigma_z = \sigma_y = \sigma_x = -\alpha E (T_1 - T)/(1 - 2\nu).$$

These equations are the basic equations for direct thermal stresses under external or internal constraints, and represent the maximum thermal stress for the particular constraint. They can refer to an entire body under external constraints or to an element of a body.

The equation in 3 defines the stress for a fully restrained body or an element. The stress is the maximum thermal stress that can be produced by a change in temperature under external or internal constraints [16].

The equation in 2 represents the maximum stress that can be produced in a body or an element restrained in two directions. An example would be the stress produced at the surface of a body under sudden heating or cooling. For steels, this would be about 200 psi/°F.

9.3. EXTERNAL CONSTRAINTS

If the temperature displacements of a body or a system of members are fully or partially prevented by outside restraints the usual method of stress computation is to find the dimensions of the freely expanded body and then to compute the forces and moments required to bring the body to the restrained state and the stresses produced by those forces.

Thus thermal problems of this type are converted into problems of conventional strength of materials. Solution depends on the availability of suitable strength formulas applicable to the problem. The temperature distribution through the body is generally uniform.

The simplest example would be a steel pipe between two fixed walls. Any change in temperature of the operating liquid in relation to atmospheric (installation) temperature will induce a change in the length of the pipe:

$$\Delta L = \alpha L (T_1 - T_{\text{atm}}).$$

If $T_{\text{atm}} < T_1$ the growth ΔL will be positive and the total force F required to restrain the pipe to the original length L will be

$$F = \alpha a E (T_1 - T_{\text{atm}}).$$

Thermal stress is then

$$\sigma = E\alpha(T_1 - T_{\text{atm}}) \quad \text{in compression}$$

where

a = the cross-sectional metal area of the pipe
α = the coefficient of thermal expansion of the pipe material.

Example 9.1. Determine the local stresses in the carbon-steel vessel shell at point B and stresses in the stainless-steel (Type 304) welded-in pan tray at point A, as shown in Fig. 9.1. At start-up the pan temperature will be the same as the temperature of the operating liquid, T_p, while the shell temperature will be only slightly above atmospheric temperature.

The stainless steel pan will expand to a greater degree than the carbon steel shell. The equation for the summation of all radial growths ΔR is as follows: (ΔR of the shell due to the internal pressure P) $-$ (ΔR of the shell due to the rise in temperature) + (ΔR of the shell due to the line load q at point B) = (ΔR of the pan due to the rise in temperature) $-$ (ΔR due to the linear force q at point B), i.e.,

$$(R/E_c)(\sigma_t - \nu\sigma_L) + \alpha_c(T_c - 70°)R + (qR^2\beta/2E_ct) = \alpha_s(T_p - 70°)R_i - (qL^3/3E_sI)$$

$$q[(R^2\beta/2E_ct) + (4L^3/E_st_p^3)] = \alpha_s(T_p - 70°)R_i - (PR^2/E_ct)[1 - (\nu/2)] = K$$

for $T_c = 70°$F, and line load q:

$$q = K/[(R^2\beta/2E_ct) + (4L^3/E_st_p^3)]$$

where $\beta = 1.285/(Rt)^{1/2}$ for $\nu = 3/10$.

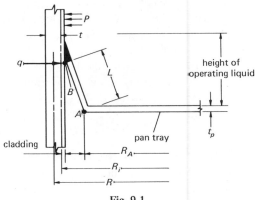

Fig. 9.1.

1. Total stress in shell at point B is:

longitudinal: $\sigma_L = [1.17q(Rt)^{1/2}/t^2] + (PR/2t)$

tangential: $\sigma_t = (qR\beta/2t) + (PR/t) = [0.64q(Rt)^{1/2}/t^2] + (PR/t)$.

2. Total stress in pan at point A is:

in tension: $\sigma_b = (6qL/t^2) + ($liquid and pan weight$/2\pi t_p R_A) < S_a$ of pan plate.

9.4. INTERNAL CONSTRAINTS

The shape of a body with nonuniform temperature distribution throughout may be such that it does not permit free expansion of the individual body elements in accordance with their local temperatures, so that stresses are produced in the body in the absence of any external constraints. This can be visualized better if a body is assumed to consist of unit elements such as cubes or fibers, each with a uniform average temperature differing from the temperature of the adjacent elements.

For example, a thick cylindrical shell can be assumed to consist of thin, mutually connected, concentric cylindrical shells. If the temperature is raised uniformly the individual shells will grow at the same rate. However, if a temperature gradient develops, for instance with heat transfer, the cylindrical elements will be at different average temperatures and expand at different rates. Since the individual cylindrical elements will be constrained by each other, thermal stresses will develop in the otherwise unrestrained shell. If the stresses become larger than the yield strength of the material at some point, residual stresses will be introduced in the shell upon return to the original temperature.

A detailed description of the general analytical methods for solving thermal stresses caused by such internal constraints is beyond the scope of this book. However, most vessel problems can be reduced to two-dimensional stresses. In simpler cases, exact analytical solutions are possible and result in closed formulas, which are sometimes cumbersome. For a practicing vessel engineer the rule is to use available closed-form results, and when necessary to simplify the problem to obtain only the most important stresses. However, the lack of derived formulas sometimes necessitates the use of approximate methods.

In the following discussion, thermal stress problems will be described briefly, and only as they occur in standard pressure vessel design practice.

Thin Plate Bar with Transverse Temperature Gradient

Figure 9.2 shows a thin, unrestrained plate bar, simply supported and subject to an uneven, one-dimensional thermal gradient across height h. Temperature T

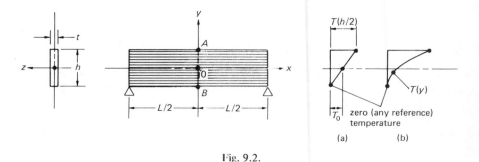

Fig. 9.2.

is defined in terms of the variable y. The stresses resulting from the uneven temperature distribution can be determined by the *equivalent thermal load method*.

To determine the thermal stresses in the bar we assume first that the bar is cut into longitudinal unit fibers of the same temperature along their lengths, which are allowed to expand freely in the x-direction. There are no shear stresses between the fibers to consider. Second, the fibers are brought into their original shapes by outside assumed forces.

The internal compressive stress produced is

$$\bar{\sigma}_x = -\alpha E T(y).$$

At this point the fibers are rejoined. Since at the ends of the actual bar there is no compressive stress $\bar{\sigma}_x$, equilibrium is restored by application of end tensile force F, distributed across bar width h at the ends. At a distance away from the end, force F will become uniformly distributed across the cross-sectional bar area:

$$F = \int_{-h/2}^{+h/2} dF \, dy = \int_{-h/2}^{+h/2} \alpha E T(y) \, t \, dy$$

$$\sigma'_x = F/ht = \frac{1}{h} \int_{-h/2}^{+h/2} \alpha E T(y) \, dy.$$

Since F is distributed non-uniformly at the ends with respect to the axis z of the bar cross section, a bending moment is introduced:

$$M_z = \sum_{-h/2}^{+h/2} (dF) y = \int_{-h/2}^{+h/2} \alpha E T(y) \, ty \, dy$$

causing the bending stress

$$\sigma_x'' = \pm \frac{M_z y}{I_z} = \frac{12y}{h^3} \int_{-h/2}^{+h/2} \alpha E T(y) y \, dy.$$

The total stress in bar away from the end in the direction of the x-axis is the summation of the above stresses: $\sigma_x = \bar{\sigma}_x + \sigma_x' \pm \sigma_x''$, or

$$\sigma_x = -\alpha E T(y) + \frac{1}{h} \int_{-h/2}^{+h/2} \alpha E T(y) \, dy \pm \frac{12y}{h^3} \int_{-h/2}^{+h/2} \alpha E T(y) y \, dy.$$

Example 9.2. Assuming a linear temperature distribution in the plate bar of Fig. 9.2, with temperature T_0 at point 0, determine the stress distribution in the bar.

$$T(y) = T_0 [1 + (2y/h)].$$

At $y = +h/2$,

$$\sigma_x = -\alpha E(2T_0) + \frac{1}{h} \left[\int_{-h/2}^{0} \alpha E T_0 \left(\frac{1 + 2y}{h} \right) dy + \int_{0}^{+h/2} \alpha E T_0 \left(\frac{1 + 2y}{h} \right) dy \right]$$

$$+ \frac{12h/2}{h^3} \left[\int_{-h/2}^{0} \alpha E T_0 \left(\frac{1 + 2y}{h} \right) y \, dy + \int_{0}^{+h/2} \alpha E T_0 \left(\frac{1 + 2y}{h} \right) y \, dy \right]$$

and the stress in the unrestrained bar at point A is

$$\sigma_x = -2\alpha E T_0 + \alpha E T_0 + \alpha E T_0 = 0.$$

This result is in accordance with the basic principle that a uniform or linearly varying temperature rise by itself does not generate thermal stresses in a solid body in the absence of constraints.

However, the linear temperature distribution in the bar will cause a deformation giving it a constant curvature:

$$(1/R) = (2\alpha T_0)/[h(1 + \alpha T_0)] \doteq (2\alpha T_0)/h.$$

If the bar is restrained in compression only $\sigma_x' = 0$ and $\sigma_x = -\alpha E T_0$. If the bar is clamped at ends, $\sigma_x'' = 0$ and again the stress is $\sigma_x = -\alpha E T_0$.

In practice, the temperature distribution will resemble to that shown in Fig. 9.2(b), generating high thermal stresses in the bar. The use of gusset plates or rib stiffeners cannot be recommended on hot vessels subject to thermal gradients of the sort which affect the bar in this example. If they are used, they must be well insulated. The interested reader will find a thorough description of the equivalent thermal load method as used to solve more general thermal problems in refs. 66, 67, 68, and 122.

Bar with Axial Temperature Variation

It can be shown by the above method that:

1. An unrestrainted bar with axial temperature variation is stress free. The total strain is

$$e_x = \int_0^L \alpha T(x)\, dx.$$

2. A restrained bar with axial gradual temperature variation $T(x)$ and average temperature T_a generates stress in the bar equal to

$$\sigma_x = -E\alpha T_a,$$

with all sections in compression.

Thick Plate with Transverse Temperature Gradient

The temperature distribution is uniaxial, the produced stress biaxial. Assuming full restraint in the x- and z-directions,

$$\sigma_x = \sigma_z = -\alpha E T(y)/(1 - \nu).$$

Then from the previously derived stress formula for a bar the following equation can be written:

$$\sigma_x = \sigma_z = -\frac{2\alpha E T(y)}{1 - \nu} + \frac{E}{h(1 - \nu)}\int_{-h/2}^{+h/2} \alpha T(y)\, dy + \frac{12yE}{h^3(1 - \nu)}\int_{-h/2}^{+h/2} \alpha T(y)\, y\, dy.$$

For a linear temperature distribution $T(y) = T_0\,[1 + (2y/h)]$ in a plane (Fig. 9.3) with edges clamped (an approximation to a thick vessel wall) at point B (the inside of a shell with a high operating temperature), the stresses are

$$\sigma_x = \sigma_z = -[2\alpha E T_0/(1 - \nu)] + [\alpha E T_0/(1 - \nu)] = -\alpha E T_0/(1 - \nu).$$

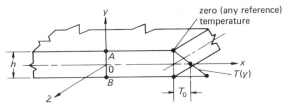

Fig. 9.3.

During a thermal shock (sudden application of hot or cold fluid of temperature T_f to the surface of the plate) the magnitude of the stress at the surface of the thick plate with an average temperature T_a and surface T_s after the contact with the fluid is given by

$$\sigma = E\alpha(T_a - T_s)/(1 - \nu) \doteq E\alpha(T_a - T_f)/(1 - \nu).$$

Uniform Hot Circular Spot in a Large, Thick Plate

If a circular hot spot of temperature T_1 with radius a develops in a shell, for instance due to a crack in the inside shell refractory, in an otherwise cold plate of temperature T (Fig. 9.4), the maximum stress is given by:

inside the hot spot: $\qquad\qquad\qquad \sigma_r = \sigma_t = -\alpha E(T_1 - T)/2;$

outside the hot spot at distance r:

$$\sigma_r = -\alpha E(T_1 - T)a^2/2r^2$$
$$\sigma_t = \alpha E(T_1 - T)a^2/2r^2.$$

These stresses are maximum stresses since any rise in the cold plate temperature will relieve them.

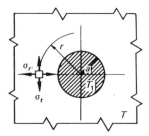

Fig. 9.4.

Thick Cylindrical Shell with Radial Temperature Distribution

For the case of a thick-wall cylindrical shell with free ends, the stresses due to a radial distribution of temperature through the wall are given by

$$\sigma_r = \frac{\alpha E}{(1 - \nu)r^2} \left[\frac{r^2 - a^2}{b^2 - a^2} \int_a^b T(r)r\,dr - \int_a^r T(r)r\,dr \right]$$

$$\sigma_t = \frac{\alpha E}{(1 - \nu)r^2} \left[\frac{r^2 - a^2}{b^2 - a^2} \int_a^b T(r)r\,dr - \int_a^r T(r)r\,dr - T(r)r^2 \right]$$

$$\sigma_L = \frac{\alpha E}{(1 - \nu)} \left[\frac{2}{b^2 - a^2} \int_a^b T(r)r\,dr - T(r) \right]$$

where

   a = inside radius (see Fig. 9.5)
   b = outside radius
   r = radius of the point under consideration
   $T(r)$ = gradient temperature of the point under consideration as a function of r. In practice determined by a heat-transfer computation.

Stresses σ_r and σ_L are algebraically additive to the pressure stresses.

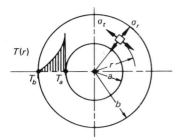

Fig. 9.5.

Assuming that the thermal gradient is linear, the gradient temperature is

$$T(r) = (b - r) T_a/(b - a).$$

At $r = b$, $T_b = 0$ or reference temperature; at $r = a$, $T = T_a$. The maximum stresses are:

at $r = a$: $\qquad\qquad\qquad \sigma_t = \sigma_L = -\frac{\alpha E T_a}{(1 - \nu)} \left[\frac{2b + a}{3(b + a)} \right]$

at $r = b$:

$$\sigma_t = \sigma_L = + \frac{\alpha E T_a}{(1 - \nu)} \left[\frac{b + 2a}{3(b + a)} \right]$$

For thin shells with $a \doteq b$, the maximum stresses are:

at $r = a$:

$$\sigma_t = \sigma_L = - \frac{\alpha E T_a}{2(1 - \nu)}$$

at $r = b$:

$$\sigma_t = \sigma_L = + \frac{\alpha E T_a}{2(1 - \nu)}$$

which is one-half the maximum thermal stress for two-dimensional constraints. The linear temperature gradient generates the smallest thermal stresses from all possible temperature distributions. For medium- and thin-wall cylinders the assumption of a linear temperature gradient is satisfactory.

Spherical Shells with Radial Temperature Gradient

In a spherical shell the principal stresses σ_t and σ_r at the point under consideration are given by

$$\sigma_r = \frac{2E\alpha}{1 - \nu} \left[\frac{r^3 - a^3}{b^3 - a^3} \frac{1}{r^3} \int_a^b T(r) r^2 \, dr - \frac{1}{r^3} \int_a^r T(r) r^2 \, dr \right]$$

$$\sigma_t = \frac{2E\alpha}{1 - \nu} \left[\frac{2r^3 + a^3}{2(b^3 - a^3)} \frac{1}{r^3} \int_a^b T(r) r^2 \, dr + \frac{1}{2r^3} \int_a^r T(r) r^2 \, dr - \frac{1}{2} T(r) \right].$$

In practice most process vessels operating at high temperatures are thermally insulated, in which case the temperature of the shell wall will be uniform and equal to the temperature of the operating liquid. However, heavy tubes in heat exchangers will be subject to temperature gradients.

Thin Cylindrical Support Skirts of Hot Vessels with Axial Thermal Gradient

In addition to the stresses produced by the vessel weight and wind or seismic moments there are thermal stresses in the cylindrical skirt supports of hot vessels generated by the temperature gradient from the skirt–shell joint down in axial direction.

The additional stresses in skirt supports will consist of the following:

1. Temperature gradient stresses will be produced by the temperature distribution alone in the cylindrical shell, regardless of whether the end is free, clamped,

or fixed. (A linear axial thermal gradient by itself does not produce any stresses in the cylinder, but it does cause expansional shell deformation.)

2. Discontinuity stresses will arise because the skirt end is welded to the shell. At the skirt–shell weld junction the skirt and shell temperatures will be nearly the same so that no radial difference in expansion occurs. However, from the joint down along the skirt any temperature gradient will cause an angular rotation of the skirt end, which is restrained by the welded joint equivalent to the clamped boundary condition of the skirt end.

3. In addition to the above stresses of thermal origin, the design vessel pressure will produce a radial deformation in the shell, which will cause stresses in unpressurized skirt.

Support skirts should be designed (a) with a thermal gradient moderated by insulation (internal and external), so that the combination of thermal and other stresses will be within the acceptable stress ranges at the welded joint, and (b) long enough to reduce the temperature difference between the bolted-down base of the skirt and the concrete base, in order to prevent any possible distortion of the skirt.

Temperature Gradient. To perform thermal stress analysis of the cylindrical support skirt of a vessel, or at least to inspect quantitatively the magnitude and effects of thermal stresses, the temperature profile of the support skirt must be known. The skirt height is usually given by process requirements, and in practice the minimum height of a support skirt is rarely below 6 ft. At this length the skirt support can be assumed to be a semi-infinite cylinder insulated inside and outside, as shown in Fig. 9.6.

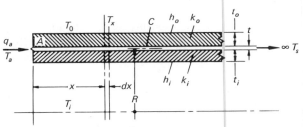

h_i, h_o = film coefficients, Btu/ft^2-hr-$^\circ$F
k_i, k_o = transverse conductivity of insulation, Btu/ft^2-hr-$^\circ$F, thickness in in.
$\quad C$ = metal conductivity, thickness in in., Btu/ft^2-hr-$^\circ$F
$\quad R_t$ = thermal metal resistance of plate to axial heat flow, hr-$^\circ$F/Btu
$\qquad$ = $144/Ct$
$\quad T_s$ = temperature of cylinder at $x = \infty$
$(T_a - T_s)$ = total temperature drop
$\quad T(x)$ = shell temperature at distance x under given ambient conditions.

Fig. 9.6. Semi-infinite cylindrical shell insulated outside and inside.

Examining a strip of unit width on such a cylindrical shell, insulated on both sides, with temperature T_a at its end and with constant temperatures T_i and T_o adjacent to the insulation, the temperature distribution is given as a function of the distance x from the end by

$$T(x) = (T_a - T_s)e^{-mx} + T_s$$

where

$$T_s = (T_i G_i + T_o G_o)/(G_i + G_o)$$
$$1/G_i = (t_i/k_i) + (1/h_i)$$
$$1/G_o = (t_o/k_o) + (1/h_o)$$
$$m = [R_t(G_i + G_o)]^{1/2}$$

At $x = 0$ the gradient becomes maximum:

$$dT(x)/dx = -m(T_a - T_s)\ °F/ft$$

This equation can be used for estimating the skirt metal temperature at the base. The maximum gradient of up to $30°F/in.$ is considered acceptable.

The analytical determination of thermal stresses due to a thermal gradient in a semi-infinite cylinder depends on the solution of the differential equation for the temperature radial growth w of the free cylinder due to the temperature distribution $T(x)$ from the end in the axial direction:

$$d^4w/dx^4 + 4\beta^4 w = -(Et\alpha/DR)T(x)$$

where

α = coefficient of thermal expansion
$D = Et^3/12(1 - \nu^2)$ flexural rigidity of the cylinder
$\beta = [3(1 - \nu^2)/R^2 t^2]^{1/4}$

and the longitudinal moment M_L and the transverse shear Q at any location:

$$M_L = -Dd^2 w/dx^2, \quad Q = -Dd^3 w/dx^3.$$

The solution and application of the results of these equations to various boundary conditions is beyond the scope of this book. The interested reader will find further information in refs. 18, 70, and 73.

Generally, it can be said that computed thermal stresses in many cases exceed the yield point, but since they consist primarily of self-limiting thermal bending

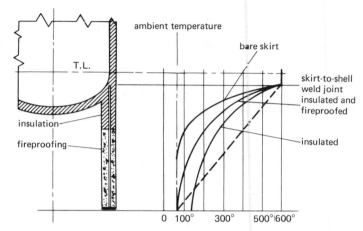

Fig. 9.7. Schematic illustration of temperature profiles under various types of insulation.

stresses they are compared rather with the allowable stress range than the allowable stresses [69, 8].

Support skirts for hot vessels are successfully designed if the temperature gradient at the skirt-shell weld joint in Fig. 9.7 is minimized by the shell insulation extending 2–3 ft below the skirt–shell weld and with the crochet space left open, as shown in Fig. 9.8.

This design is satisfactory even for large-diameter vessels with considerable radial expansion. However, some difficulties such as weld cracking have been encountered on large-diameter vessels in cyclic thermal service accompanied sometimes by thermal shock. For such vessels a support skirt lapped to the

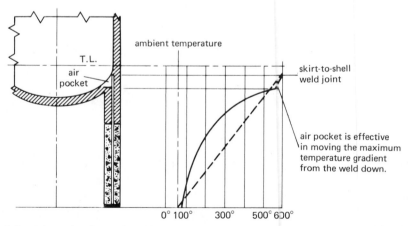

Fig. 9.8. Schematic illustration of temperature profile for insulated and fireproofed skirt with air pocket.

bottom cylindrical shell portion would seem to be a more suitable design for even and better heat transfer. A more involved analysis becomes necessary here. Short, longitudinal slots located below the weld joint have not proven completely successful in practice for relieving thermal stresses, since they weaken the shell and behave as discontiuities [74, 75].

9.5. THERMAL STRESS RATCHET UNDER STEADY MECHANICAL LOAD

Application of a steady mechanical load to a system subject to a cyclic operating temperature may result *under certain conditions* in cycling of combined thermal and mechanical stresses and a progressive increase in the plastic strain in the entire system. This action of cyclic progressive yielding is called *thermal ratcheting* and may lead to large distortions of the system and ultimately to failure.

The basic explanation of this phenomenon is given in the following descriptive example. A rigid slab M of weight W (Fig. 9.9) is supported by metal cylinder A with cross-sectional area a and bar B with cross-sectional area b. It is assumed that both materials behave in an elastic–perfectly plastic manner.

First Half of the Thermal Cycle

The initial stress σ_i induced in the cylinder and the bar, both of uniform temperature T, is

$$\sigma_i = W/(a + b)$$

and is given by point $A \equiv B$ on the stress–strain curve in Fig. 9.10.

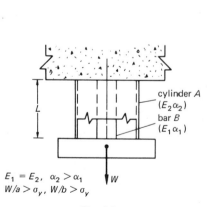

$E_1 = E_2$, $\alpha_2 > \alpha_1$
$W/a > \sigma_y$, $W/b > \sigma_y$

Fig. 9.9.

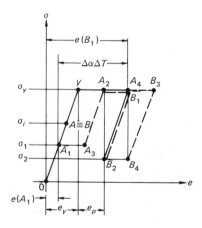

Fig. 9.10.

If the temperature of the system is raised to T_1, the free length L of the cylinder increases by $\Delta L = (\alpha_2 - \alpha_1) L (T_1 - T)$ over the free length of the bar, and a portion of the load carried by the cylinder will be transferred to the bar. The stress in the cylinder will drop to the point A_1 on the curve, and if the increase in temperature is large enough, yield stress σ_y will be induced in the bar B. The produced strain is given by the point B_1, where the cylinder takes over a part of the load to prevent further yielding of the bar and the force equilibrium in the system is restored.

The tensile stress at point A_1 in the cylinder is given by

$$a\sigma_1 = \sigma_i(a + b) - \sigma_y b.$$

The strains in the cylinder and the bar produced by weight W differ only by the amount $\Delta\alpha\Delta T$, or

$$e(B_1) = e(A_1) + \Delta\alpha\Delta T$$

Since the system forms a continuous structure the total increase in the common length must be the same for the cylinder and the bar. However, the part of the bar elongation $e(B_1)$ is now permanent.

Second Half of the Thermal Cycle

The system is cooled to initial temperature T and the situation is now reversed. The free length of the bar has been increased by the permanent plastic elongation and the bar unloads a portion of its load along the line $B_1 = B_2$. The cylinder takes over this part of the load. Assuming area a is such that yield stress is induced, the cylinder deforms partially plastically to point A_2. However, no plastic deformation occurs in the bar.

Since the final lengths of the bar and the cylinder must be the same, point A_2 lies on the same vertical line as point B_2. The tensile stress σ_2 in the bar at point B_2 is given by

$$b\sigma_2 = \sigma_i(a + b) - \sigma_y a.$$

At the end of the next thermal cycle the final state of stress–strain will shift to point B_4 in the bar and to point A_4 in the cylinder, as indicated by dashed lines in Fig. 9.10.

The lengths of the bar and the cylinder are permanently increased in each cycle and the permanent elongation e_p at the end of the cycle will be, from the stress–strain diagram,

$$e_p = \Delta\alpha\Delta T + e(A_1) - e(Y) - [e(B_1) - e(B_2)]$$
$$= \Delta\alpha\Delta T - [e(Y) - e(A_1)] - [e(B_1) - e(B_2)]$$
$$= \Delta\alpha\Delta T - [(\sigma_y - \sigma_1)/E] - [(\sigma_y - \sigma_2)/E].$$

Substituting for σ_1 and σ_2 in the above equation, we get

$$e_p = \Delta\alpha\Delta T - (\sigma_y/E) + [\sigma_i(a + b)/aE] - (\sigma_y b/aE) - (\sigma_y/E)$$
$$+ [\sigma_i(a + b)/bE] - (\sigma_y a/bE)$$
$$= \Delta\alpha\Delta T - [(\sigma_y - \sigma_i)/E][2 + (b/a) + (a/b)].$$

If $\sigma_y - \sigma_i$ is very small and can be disregarded, e_p is maximum and equal to $\Delta\alpha\Delta T$.

It is important to note that the above system acts as a continuous structure. Therefore ratcheting can not develop in joints like the flange–bolt joint. A practical example of thermal ratcheting would be the cyclic growth of a thick cylindrical shell of uniform material subject to internal pressure and fluctuating operating temperature which develops a radial thermal gradient through the wall. The design rules of Code Division 2 should then be applied.

9.6. DESIGN CONSIDERATIONS

In the design of pressure vessels the designer tries to eliminate or at least minimize thermal stresses as much as possible. Several methods in use are discussed here.

1. One of the simplest and effective ways to avoid external constraints is by using *floating construction*, which *eliminates external constraints*. An example of floating construction is shown in Fig. 9.11, where a heat exchanger is attached to a process column by two brackets. The top bracket carries the entire weight of the heat exchanger, while the bottom bracket with slotted holes supports the heat exchanger horizontally and provides for the vertical differential thermal expansion. The nuts on bolts in the bottom bracket are hand tight and secured by tack welding. Another example are the horizontal and vertical drums on sliding support plates.

2. If external constraints cannot be entirely eliminated, *local flexibility* capable of absorbing the expansion is provided in the form of expansion joints or a flexible structural member. An example would be the pan-to-shell joint shown in Fig. 9.1, which can absorb the radial growth of the pan without

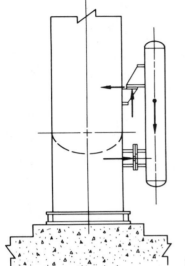

Fig. 9.11.

causing the buckling that would occur if the pan were welded directly to the shell.

3. Sometimes a *change in design* is desirable to solve a thermal problem. For instance, in the case of a large U-tube heat exchanger too large a difference between the in and out temperatures on the tube side caused the tube sheet to warp and a leakage occurred. The suggested remedy was to replace U-tube design with once-through flow construction.

4. Shapes of weldments or castings that might cause a steep temperature gradient under operating conditions can be avoided by proper *contouring* of the part. Sources of stress concentration, holes, abrupt changes in cross-sections, or mass accumulation should be minimized. Thermal expansion can be better handled by dividing large parts.

5. *Selection of proper material* or combination of materials can minimize thermal stress. Consider the bolted, flanged joint in high-temperature service shown in Fig. 9.12.

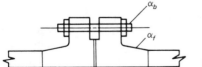

Fig. 9.12.

The thermal expansion coefficient α_b and α_f of the bolt and flange, respectively, should be approximately the same. If $\alpha_b > \alpha_f$, leakage of the joint occurs at high operating temperatures and if the bolt is tightened a permanent set occurs when the joint returns to normal temperatures. If $\alpha_f > \alpha_b$ the bolt becomes overstressed and permanently stretched at high temperatures. On return to lower temperatures the joint leaks.

6. A temperature gradient can be moderated by *selective use of insulation*, as was done by insulating the top of the support skirts for hot vessels in Fig. 9.8.

10
Weld Design

10.1. INTRODUCTION

Today, welding is the most commonly used method of fabrication of pressure vessel shells. Structural, nonpressure parts such as stiffening rings, lifting lugs, support clips for piping, internal trays, and other equipment are usually attached to vessel wall by welding as well. Welded joints, instead of bolted joints, are sometimes used for piping-to-vessel connections to obtain optimum leak-proof design where desirable, as with lethal fluids or in high-temperature service.

A structure whose component parts are joined by welding is called a *weldment*. A *weld* is defined as a localized union of metal achieved in plastic and molten states, with or without the application of pressure and additional filler metal. Basically, there are three welding methods:

1. Forge welding is the oldest method, applicable to low-carbon steel. The joint is not particularly strong.

2. Fusion welding is a process that does not require any pressure to form the weld. The seam to be welded is heated, usually by burning gas or an electric arc to fusion temperature, and additional metal, if required, is supplied by melting a filler rod of suitable composition.

3. Pressure welding is used in processes such as resistance welding, which utilizes the heat created by an electric current passing against high resistance through the two pieces at the contact interface.

The most widely used industrial welding method is *arc welding*, which is any of several fusion welding processes wherein the heat of fusion is generated by an electric arc.

Residual stresses in a weld and in the region adjacent to a weld are unavoidable and complex, but they are not considered dangerous when static loads are applied. If the weld residual stress is superposed on the stress caused by an external load and exceeds the yield point of the material, a small amount of local plastic yielding will redistribute the stress. This is particularly true for ductile materials, and one important requirement for a good weld is high ductility. In order to prevent loss of ductility in the heat-affected zones, only low-carbon, nonharden-

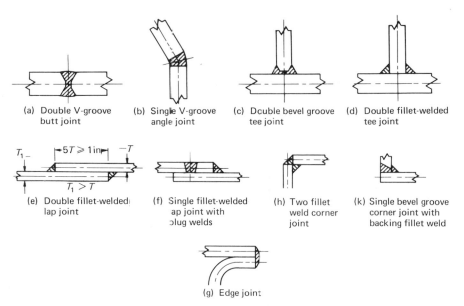

(a) Double V-groove butt joint

(b) Single V-groove angle joint

(c) Double bevel groove tee joint

(d) Double fillet-welded tee joint

(e) Double fillet-welded lap joint

(f) Single fillet-welded lap joint with plug welds

(h) Two fillet weld corner joint

(k) Single bevel groove corner joint with backing fillet weld

(g) Edge joint

Fig. 10.1.   Some forms of welded joints in combination with different weld types.

able steels with less than 0.35 percent carbon content are used as construction material. Carbon is an effective steel hardener; however, with the usual manganese content of 0.30–0.80 percent it does not cause any difficulties when present in steel up to 0.30 percent. In addition, the residual stresses in heavier plates are usually removed by post-weld heat treatment. Cooling after welding causes dimensional changes in the weld due to temperature reduction and phase change; these changes may result in occasional cracking in the weld or the heat-affected zones. Fortunately, the steel lattice expands as it becomes ferritic at lower temperature, but contracts because of the reduction in temperature.

For design purposes, *welds* can be divided into three basic types, calling for different design methods and different design stresses: *groove, fillet, and plug welds. Welded joints* are described by the position of pieces to be joined, and are usually divided into five basic types: *butt, tee, lap, corner, and edge.*

The types of joints should never be confused with the types of welds. In Fig. 10.1 the basic types of welded joint in combination with different types of welds are shown. For a designer, the immediate task is to select the joint design and the type of attachment weld, and to compute its size.

10.2.   GROOVE WELDS

Groove welds can be subdivided according to the edge conditions, as shown in Table 10.1. All types of grooves shown can be used in butt, tee, or corner weld

Table 10.1.   Types of Groove Welds.

	SINGLE	DOUBLE
Square		
Bevel groove		
V groove		
J groove		
U groove		

joints. Butt or tee joints with V or double-bevel groove welds are used extensively, particularly where cyclic loads are expected. They are easier to design than fillet welds, but harder to fabricate. The edge preparation required depends on the plate thickness and is usually specified by the applicable code. Detailed weld bevel dimensions may differ for shop and field welds and are subject to the approval of the welding engineer. A careful fit-up is required for butt joints, but a slight mismatch may be permissible. In the absence of specifications, ref. 82 can serve as a guide. Edge beveling or grooving primarily aids weld penetration. Some welding processes (e.g., automatic submerged arc welding) are capable of very deep penetration, while others (e.g., manual arc welding) are not.

The strength of a groove weld is based on the cross-sectional area subject to shear, tension, or compression and the allowable stress for the weld metal, which is the same or nearly the same as the allowable stress for the base metal. Stresses in groove welds are computed from standard formulas for stresses in tension, bending, or shear and combined in the standard manner. The complete penetration groove is the most reliable of all weld types. There is no significant stress concentration due to weld shape, since the force lines are smooth and continuous. Joint efficiency depends on the type of weld examination, the welding procedure, and type of load. It is usually given in applicable codes and ranges between 0.80 and 1.00.

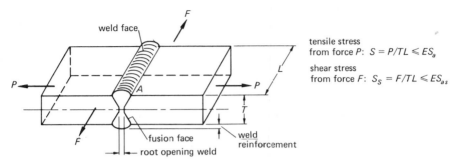

tensile stress
from force P: $S = P/TL \leqslant ES_a$

shear stress
from force F: $S_S = F/TL \leqslant ES_{as}$

S_a = allowable tensile or compressive stress of the base metal
S_{as} = allowable shear stress of the base metal
E = joint efficiency for the groove welds
T = width of the groove weld equal to the thickness of welded plate at joint
L = length of the welded joint

Fig. 10.2. Full-penetration double V-groove butt joint.

Determining Groove Weld Size

In Fig. 10.2 a typical butt joint is shown (both welded parts are in the same plane) with full-penetration double V-groove weld. It can be seen that point A where the weld reinforcement reenters the base metal is a source of some stress concentration and potential failure under repetitive load. Removal of the weld reinforcement, by grinding flush, increases the fatigue strength of the weld and is therefore required by some codes.

In Fig. 10.3 a butt joint with incomplete-penetration groove welds is illustrated. If the weld metal is not removed by gouging on the root before welding the other side, or if the depth of penetration is not full, the welded joint becomes an incomplete-penetration weld with a lower joint efficiency. Because of its poor strength under cyclic operating conditions this weld type is used mainly for connecting nonpressure parts to the vessel shell under static loads where fillet welds would be too large. The resisting weld area equals to $(h_1 + h_2)L$.

In Fig. 10.4 a tee joint with a full-penetration double-bevel groove weld plus fillet weld on each side is shown. The throats of the fillet welds may be included in computation of the critical stress-resisting area. However, the main purpose

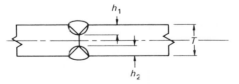

Fig. 10.3. Butt weld with incomplete-penetration double V-groove weld.

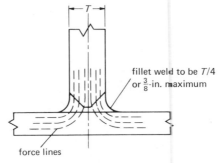

Fig. 10.4.   Tee joint with double-bevel groove weld.

of the fillet welds here is to reduce the stress concentration, so the resisting area is equal to T times L.

To summarize, in order to determine the maximum stresses in a groove weld of the width T the standard stress formulas are used with properties of the weld critical section (resisting area a, section modulus Z and polar modulus J). The different computed stress components at the same location are combined analytically or by the Mohr circle, and the maximum shear and principal stresses are evaluated. They are compared with the allowable weld stresses based on the allowable stresses of the base metal times the weld joint efficiency.

Example 10.1. Check whether the size of the weld connecting the lifting lug to the vessel wall in Fig. 10.5 is adequate. Use the impact factor 2 for sudden loads. Assume the maximum allowable stress for the groove weld to be equal to the strength of the lifting lug material.

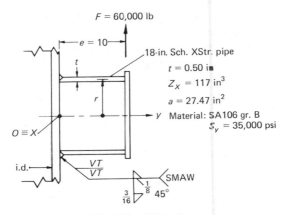

Fig. 10.5.   Lifting lug.

Bending stress: $S_b = M/Z_x = 2Fe/\pi r^2 t$
$= 2 \times 60,000 \times (10/117) = 10,260$ psi.

Shear stress: $S_s = 2F/a = 2 \times 60,000/27.5 = 4,370$ psi.

Combined maximum shear:

$$S_s' = [(S_b/2)^2 + S_s^2]^{1/2} = [(10,260/2)^2 + 4,370^2]^{1/2} = 6,740 \text{ psi.}$$

Combined maximum tension:

$$S = (S_b/2) + [(S_b/2)^2 + S_s^2]^{1/2} = (10,260/2) + 6,740 = 11,870 \text{ psi.}$$

Safety factors:

in shear: S.F. $= 0.6 \times 35,000/6,740 = 3.12$

in tension: S.F. $= 35,000/11,870 = 2.95$

Both safety factors are acceptable.

Note. Both maximum stresses in shear and tension must be smaller than the allowable design weld stresses in tension and shear for weld. Also the stresses in the shell have to be checked.

10.3. FILLET WELDS

A fillet weld is a weld with an approximately triangular cross section, joining two surfaces at right angles. The size of a fillet weld is specified by the leg length *w* (see Fig. 10.6) of its largest inscribed right triangle (in Europe, by its throat thickness *t*). The 45-degree fillet weld with equal legs is the most economical and therefore generally used. Since no edge preparation is required the cost is low, but so is the allowable stress. Stress concentrations at the root or at the toe can cause failure under variable loads. The main weld definitions are shown in Fig. 10.6. In stress computations no credit can be taken for the weld reinforcement. A one-sided fillet weld in tee or lap joints should be avoided because of its very low static and fatigue strengths. The face of the fillet weld may vary from convex to concave.

Stresses induced in fillet-welded joint are complex because of the eccentricity of the applied load, the weld shape, and notch effects. They consist of shear, tension, and compression stresses. Distribution of stresses is not uniform across the throat and leg of a fillet weld and varies along the weld length as well. In Fig. 10.7 an approximate stress distribution in a transverse weld is shown. A large stress concentration is apparent at the root of the fillet weld. Under such conditions it becomes nearly impossible to obtain correct maximum analytical

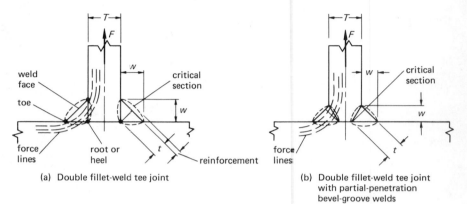

(a) Double fillet-weld tee joint

(b) Double fillet-weld tee joint with partial-penetration bevel-groove welds

T = welded part thickness

w = weld leg

t = weld throat, the shortest distance from root to face of the fillet weld. For a 45-degree weld, $t = 0.707w$.

Fig. 10.6. Definitions relating to fillet welds.

stress formulas. For practical reasons it is important to have a simple method of stress analysis to compute the proper amount of the fillet weld and to provide adequate strength for all types of connections for static or fluctuating loads. However, excessive welding may be a major factor contributing to an increase in welding cost, residual stresses, and distortions.

In order to determine the nominal (average) stresses and the size the fillet welds from standard tension, bending, and torsion stress formulas, weld section properties such as area, section modulus, and polar modulus must be computed. These properties of the critical sections can be computed by treating the critical section of the weld in two ways: as an area or as a line.

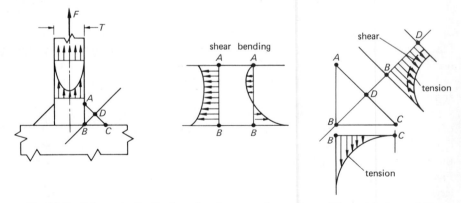

Fig. 10.7. Schematic distribution of main stresses in transverse fillet weld of a tee joint.

Determining the Size of a Fillet Weld by Treating the Critical Section as an Area

Basically, there are two directions of applied loads on the fillet welds: *parallel* to the weld length and *transverse* to the weld length.

The average shear and normal stresses have been computed for both cases using simplified freebody diagrams (Table 10.2). The simplified approach gives different maximum (average) stress values for transverse and parallel welds. Further difficulties arise when both weld types are used in combination. Welded connections in practice consist of a group of welds subjected to one or more types of loadings simultaneously. A number of simplifying assumptions have to be made to arrive at practical, unified design formulas applicable to both types of welds and giving safe welded connections when compared with the test results.

1. The critical resisting area for both types of welds or their combinations is the minimum cross-sectional area at the throat plane and equal to the weld throat thickness times the effective weld length. It disregards the possibility that another section can actually be subject to higher combined stress. The exception is Code Division 1, where the leg size is used instead of the throat in computations of the critical area of a fillet weld. However, the lower joint efficiency compensates for the increased area in this case.

2. The nominal stresses are computed by substituting the properties of this weld area (area *a*, section modulus *Z*, or polar section modulus *J*) into the standard stress formulas for tension, bending, or torsion.

3. If several stresses in different planes must be combined at the same location to determine the maximum stresses, the conventional analytical formula for determining the combined maximum stress can be used, as was done for the groove welds. However, it can be seen even from the simplified diagrams in Table 10.2 that the maximum induced stresses (normal or shear in fillet welds) are hard to define and use in formulas for combined stress. Therefore, in the case of fillet welds an alternative procedure is to treat any stress on the weld throat, regardless of direction, as a shear stress. If a fillet weld is then subject to combination of stresses S_1, S_2, S_3 (shear or tensile) acting at right angles to each other, the resultant stress is considered as a vector sum of all three stresses:

$$S = (S_1^2 + S_2^2 + S_3^2)^{1/2}.$$

All stresses to be combined must occur at the same location. This totally arbitrary approach lacks analytical justification, but it gives conservative weld sizes in agreement with tests and is frequently used for simplicity. It is left up to the designer which approach he prefers to use.

4. The maximum resulting stresses are limited by the allowable stress of the base metal, assuming the weld metal to be stronger, times the joint efficiency for the fillet welds. A transversely stressed fillet weld can sustain higher loads

Table 10.2 Simplified Stress Analysis of Double Fillet Welds

T-JOINT WITH PARALLEL DOUBLE FILLET WELDS	*T*-JOINT WITH TRANSVERSE DOUBLE FILLET WELDS
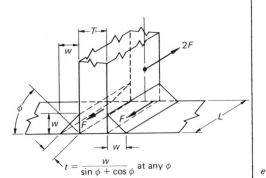	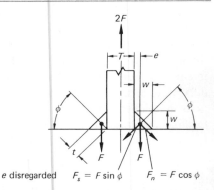
$t = \dfrac{w}{\sin \phi + \cos \phi}$ at any ϕ	e disregarded $F_s = F \sin \phi$ $F_n = F \cos \phi$

Average shear stress at angle ϕ cutting plane:

$$S_s = \frac{F}{tL} = \frac{F(\sin \phi + \cos \phi)}{wL}$$

Maximum average shear stress at 45-degree plane:

$$S_s = \frac{F}{tL} = \frac{1.414F}{wL}$$

where

F = applied force per weld
L = effective weld length
w = weld leg
$t = 0.707w$ weld throat at 45-degree plane

Average normal stress S at any angle ϕ:

$$S_n = \frac{F \cos \phi}{tL} = \frac{(F \cos \phi)(\cos \phi + \sin \phi)}{wL}$$

Average shear stress S_S at any angle ϕ:

$$S_S = \frac{F \sin \phi}{tL} = \frac{(F \sin \phi)(\cos \phi + \sin \phi)}{wL}$$

Maximum average shear stress occurs at $\phi = 67.5$ degrees:

$$\text{max. } S_S = \frac{1.2F}{wL} \text{ with } S_n = \frac{0.5F}{wL}$$

max. combined

$$S_S = \left[\left(\frac{S_n}{2} \right)^2 + S^2 \right]^{1/2} = \frac{1.21F}{wL}$$

Maximum average normal stress occurs at $\phi = 22.5$ degrees:

$$\text{max. } S_n = \frac{1.2F}{wL} \text{ with } S_S = \frac{0.5F}{wL}$$

Average normal and shear stresses at $\phi = 45$ degrees:

$$S_n = \frac{F}{wL}, \quad S_S = \frac{F}{wL}$$

$$\text{max. combined } S_n = 1.618 \frac{F}{wL}$$

$$\text{max. combined } S_S = 1.12 \frac{L}{wL}$$

than one stressed longitudinally (parallel). Some specifications allow the use of a higher allowable stress for transverse welds. However, if transverse welds are used in combination with parallel welds, the lower allowable stress for parallel welds must be used.

5. When parallel and transverse fillet welds are used together it is assumed that the load is uniformly distributed between the welds. Distribution of the shear stress is uniform throughout the weld sections. Bending stress is linearly distributed according to beam theory. (If two or more types of welds—groove, fillet, or plug—are combined in a single joint, the allowable capacity of each is computed with reference to the group to determine the capacity of the entire group of welds.)

6. The moment of inertia about the longitudinal axis of the weld cross section is disregarded.

7. Stresses computed by stress formulas are nominal, due to external loads only. The effects of stress concentrations, residual stresses, and shape of weld are neglected for static loads.

8. The effect of possible deflections of the parts connected by welding on the stresses in the weld is neglected. The parts are assumed to be rigid.

These assumptions result in conservative weld sizes. Considering the possible variations in weld quality, however, they seem to be justified.

A considerable disadvantage of this method is that the throat size t of the fillet weld has to be first assumed for stress computations. If later the maximum stress is too far from allowable stress the computations must be revised. However, this procedure is very suitable for checking existing welds.

Example 10.2. A structural clip with double fillet weld is subjected to forces F and P as shown in Fig. 10.8. Consider the effects of both forces separately and combined. Joint efficiency E is based on the weld throat width t.

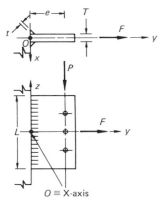

Fig. 10.8.

(a) Critical weld cross-section properties.

Area: $\qquad a = 2tL = 1.414wL.$

Section modulus: $\qquad Z_x = 0.707wL^2/3.$

(b) Stresses from force P.

Bending: $\qquad S_1 = M/Z_x = Pe/Z_x.$

Direct shear: $\qquad S_2 = P/2tL = P/1.414wL.$

$$\text{maximum } S = (S_1^2 + S_2^2)^{1/2} \leqslant ES_a.$$

(c) Stresses from force F.

Shear: $\qquad S_3 = F/2tL = F/1.414wL \leqslant ES_a.$

(d) Maximum combined stress due to forces F and P.

$$\text{maximum } S = [(S_1 + S_3)^2 + S_2^2]^{1/2} \leqslant ES_a.$$

Notes. The whole load is transmitted through the fillet welds only and no credit is taken for possible bearing between the lower part of the clip and the vessel wall. The assumption that the shear S_2 is carried uniformly in welds is generally accepted.

Example 10.3. A thin support clip of a length L with fillet welds is subjected to force F, as shown in Fig. 10.9. Determine the stresses in the welds. Joint efficiency E is based on the weld leg w.

Bending stress: $\qquad S_1 = q/wL \qquad$ where $\qquad q = Fe/(w + T).$

Shear stress: $\qquad S_2 = F/2wL.$

Combined maximum stress: $\qquad S = (S_1^2 + S_2^2)^{1/2} \leqslant ES_a.$

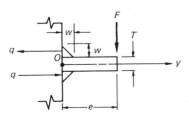

Fig. 10.9.

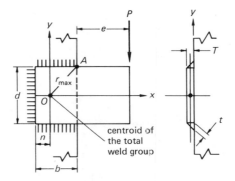

Fig. 10.10.   Structural clip subjected to an eccentric twisting load P.

Example 10.4. Determine the maximum stress in the fillet weld around the clip shown in Fig. 10.10.

(a) Critical weld cross-section properties.

Throat area: $a = 2bt + dt$.

The procedure for determining the polar moment of inertia J of the total weld area of a weld joint is as follows. (1) The polar moment of inertia J_0 of each straight section of the weld area with reference to its own centroid and to the centroid of the entire weld is computed

$$J_0 = (Lt^3/12) + (L^3t/12) + (Lt)\, r_c^2$$

and disregarding $Lt^3/12$

$$J_0 = Lt[(L^2/12) + r_c^2]$$

where r_c is the distance from the origin O to the centroid of the weld section (see Fig. 10.11). The formula is valid for both positions. (2) The polar moment of inertia J of the entire weld is the summation of all J_0's of the individual weld section areas. In Fig. 10.12 the total moment J is

$$J = 2bt[(b^2/12) + r_c^2] + dt[(d^2/12) + n^2]$$

where

$$r_c = (x_c^2 + y_c^2)^{1/2}$$
$$n = (2bt)(b/2)/(2bt + dt) = b^2/(2b + d).$$

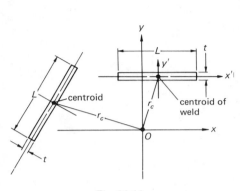

Fig. 10.11.

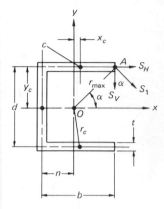

Fig. 10.12.

(b) The maximum torsional stress S_1 is at point A:

$$S_1 = P(e + b - n) r_{max}/J \quad \text{where} \quad r_{max} = [(b - n)^2 + (d/2)^2]^{1/2}.$$

This stress can be resolved into a vertical component

$$S_V = S_1 \cos \alpha$$

and a horizontal component

$$S_H = S_1 \sin \alpha.$$

(c) The direct shear stress is

$$S_2 = P/(2bt + dt).$$

S_2 is vertical and additive to S_V.

(d) The maximum shear stress at point A combined is

$$S = [(S_2 + S_V)^2 + S_H^2]^{1/2} \leqslant ES_a.$$

Treating the Fillet Weld Critical Section as a Line

This method is simpler and more direct than the previous one. It permits finding the weld size directly. Actually, it originated from the "weld area method" by using linear moments of inertia, polar and rectangular, of the weld outlines in the standard design stress formulas. It can be shown that the properties J and I

of thin areas are equal to the properties J_w and I_w of a weld section treated as a line, multiplied by its uniform thickness t, with negligible error; for instance $Z = Z_w t$.

Instead of unit stress S in psi the result in standard design formulas is unit force $f = S \times t$ in lb/lin. in. of weld length. If several forces f in different planes are acting at the same point they are combined vectorially. This gives a clearer and mentally more acceptable picture than does the previous method. The linear moments J_w and I_w and the linear section modulus Z_w of a weld outline can easily be computed by using the parallel-axis theorem or by direct integration. For convenience the J_w and Z_w values of some commonly used weld outlines are given in Table 10.3.

The required weld leg w is then determined by dividing the resultant computed unit force f by the allowable unit force f_w. The entire design procedure to compute the resultant f can be summarized as follows:

1. Draw a freebody diagram of the attachment to be connected with a fillet weld and all acting forces and moments.

2. Draw the proposed weld outline and select orthogonal reference axes x, y, z, usually axes of symmetry with the reference point O in the centroid of the weld outline. Compute the linear properties of the total weld outline (length L, J_w, and Z_w, as required).

3. Determine all forces and moments with reference to the weld outline. If required, they are resolved into components.

4. Using conventional stress formulas with linear properties, the maximum unit forces f_x, f_y, f_z due to all external loads are computed and, if at the same location, they are vectorially added in the resultant force f:

$$f = (f_x^2 + f_y^2 + f_z^2)^{1/2} .$$

5. The maximum allowable unit force f_w in lb/lin. in. on an equal-leg fillet weld w equal to one inch is given by

$$f_w = 0.707 E S_a$$

where S_a is the maximum allowable stress of the base metal and E is the joint efficiency of the fillet weld, based on throat area. If E is based on the weld leg area, then $f_w = E S_a$.

6. The required minimum weld leg is then

$$w = f/f_w \text{ (in.).}$$

The disadvantages of the above method is that it cannot readily be applied to combinations of fillet welds with different throat thicknesses, and possible error

Table 10.3. Linear Z_w and J_w of Compound Weld Sections.

OUTLINE OF THE WELDED JOINT	SECTION MODULUS Z_w ABOUT AXIS x-x (in.^{2})	POLAR MODULUS J_w ABOUT CENTROID O (in.^{3})
1.	$Z_w = \dfrac{d^2}{6}$	$J_w = \dfrac{d^3}{12}$
2.	$Z_w = \dfrac{d^2}{3}$	$J_w = \dfrac{d^3 + 3db^2}{6}$
3.	$Z_w = bd$	$J_w = \dfrac{b^3 + 3bd^2}{6}$
4.	$Z_w = bd + \dfrac{d^2}{3}$	$J_w = \dfrac{(b+d)^3}{6}$
5. $n = \dfrac{b^2}{2b+d}$	$Z_w = bd + \dfrac{d^2}{6}$	$J_w = \dfrac{(2b+d)^3}{12} - \dfrac{b^2(b+d)^2}{(2b+d)}$
6. $e = \dfrac{d^2}{b+2d}$	$Z_w = \dfrac{2bd + d^2}{3}$ $Z_w = \dfrac{d^2(2b+d)}{3(b+d)}$	$J_w = \dfrac{(b+2d)^3}{12} - \dfrac{d^2(b+d)^2}{(b+2d)}$
7. $e = \dfrac{d^2}{b+2d}$	$Z_w = \dfrac{2bd + d^2}{3}$ $Z_w = \dfrac{d^2(2b+d)}{3(b+d)}$	$(1)\,J_w = \dfrac{(b+2d)^3}{12} - \dfrac{d^2(b+d)^2}{(b+2d)}$

Table 10.3. (*Continued*)

OUTLINE OF THE WELDED JOINT	SECTION MODULUS Z_w ABOUT AXIS x-x (in.^{2})	POLAR MODULUS J_w ABOUT CENTROID O (in.^{3})
8.	$Z_w = \dfrac{4bd + d^2}{3}$ $Z_w = \dfrac{4bd^2 + d^3}{6b + 3d}$	$^{(2)}J_w = \dfrac{d^3(4b + d)}{6(b + d)} + \dfrac{b^3}{6}$
9.	$Z_w = bd + \dfrac{d^2}{3}$	$^{(1)}J_w = \dfrac{b^3 + 3bd^2 + d^3}{6}$
10.	$Z_w = 2bd + \dfrac{d^2}{3}$	$^{(2)}J_w = \dfrac{2b^3 + 6bd^2 + d^3}{6}$
11.	$Z_w = \dfrac{\pi d^2}{4} = \pi r^2$	$J_w = \dfrac{\pi d^3}{4} = \dfrac{\pi r^3}{32}$

[1] Distance between vertical welds negligible.
[2] Distance between vertical (horizontal) parallel welds negligible.

in the computation of I_w for welds of high ratios of t to the depth of the weld outline. However, it shortens design time, is of acceptable accuracy, and its use is recommended here.

Example 10.5. Determine the size of the fillet weld for the plate support clip in Fig. 10.13, $S_a = 12,000$ psi for base metal, $E = 0.55$ based on the weld leg.
 (a) Weld properties:

Weld length: $\qquad L = 2d = 6 \times 2 = 12$ in.

Linear section modulus: $\quad Z_x = 2d^2/6 = 12$ in.^{2}

 (b) Unit forces due to V.

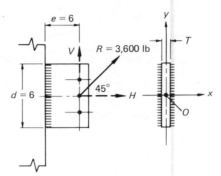

Fig. 10.13.

Bending: $f_1 = Ve/Z_x = 2,550 \times 6/12 = 1,275$ lb/in.

Shear: $f_2 = V/L = 2,550/12 = 215$ lb/in.

(c) Unit force due to H.

Shear: $f_3 = H/2L = 2,550/12 = 215$ lb/in.

(d) Weld size due to combined forces V and H:

$$f = [(f_1 + f_3)^2 + f_2^2]^{1/2} = [(215 + 1,275)^2 + 215^2]^{1/2} = 1,505 \text{ lb/in.}$$

Leg size: $w = f/f_w = 1,505/(0.55 \times 12,000) = 0.228$ in.

Use $\frac{1}{4}$-in. fillet weld all around. Corrosion allowance is added only if the vessel is not painted or exposed to a corrosive environment.

Primary and Secondary Fillet Welds

According to the importance and the magnitude of the load transmitted, fillet welds can be subdivided into primary and secondary.

Primary welds carry the entire load in some way, and if they fail the structure fails. The welds must be as strong as the other members of the structure. An example is the weld connecting a support skirt to a vertical vessel subject to external loads. All such welds are generally designed as continuous.

Secondary welds, on the other hand, are welds subject to comparatively light loads or no loads at all, only holding the structural parts together. A good example are the vertical welds on built-up saddles for horizontal drums. Obviously, they need not be continuous or excessively large. However, AWS Specifications [79] recommend a minimum fillet weld size based on the thicker plate,

Table 10.4. Recommended Minimum Weld Sizes for Thick Plates (AWS).

THICKNESS T OF THICKER PLATE WELDED (in.)	MINIMUM LEG SIZE OF FILLET WELD w (in.)
$T \leqslant \frac{1}{2}$	$\frac{3}{16}$
$\frac{1}{2} < T \leqslant \frac{3}{4}$	$\frac{1}{4}$
$\frac{3}{4} < T \leqslant 1\frac{1}{2}$	$\frac{5}{16}$
$1\frac{1}{2} < T \leqslant 2\frac{1}{4}$	$\frac{3}{8}$
$2\frac{1}{4} < T \leqslant 6$	$\frac{1}{2}$
$6 < T$	$\frac{5}{8}$

as given in Table 10.4. With a minimum size given the cost of welding can be reduced by using intermittent welds. The size of the intermittent weld can be computed from the condition that the strength or the areas of both (continuous and intermittent) must be equal:

$$w_c L = w_i(xL) \quad \text{and} \quad x = (w_c/w_i)\,100$$

in percent of length L of the continuous weld, where w_c is the leg of the required continuous weld and w_i is the leg of the required intermittent weld. The recommended lengths and spacings of the intermittent welds of leg size w_i expressed as a percentage of w_c are given in Table 10.5. All fillet welds for structural nonpressure attachments, inside and also outside, welded to the pressure parts, where an occasional inspection for corrosion is difficult, should be continuous or with intermittent strength welds and seal welds between. The minimum length of an intermittent fillet weld is $1\frac{1}{2}$ in. or 4 times plate thickness, whichever is larger. The minimum practical size of the weld leg w is $\frac{3}{16}$ in. used for strength welds, and $\frac{1}{8}$ in. for seal welds.

Example 10.6. If $w_c = \frac{3}{16}$ in. and the required minimum weld leg size is $w = \frac{5}{16}$ in., the required length of the intermittent weld w_i in percent of the length of the

Table 10.5. Intermittent Welds.

PERCENT OF LENGTH OF CONTINUOUS WELDS	LENGTH–SPACING FOR INTERMITTENT WELDS
60	3–5
50	2–4 3–6 4–8
40	2–5 4–10
30	3–10
25	2–8 3–12

continuous weld w_c is $x = (3/5)100 = 60$ percent. Use $\frac{5}{16}$-in. weld, 3 in. long on 5-in. centers with 2-in. gap between the welds.

10.4. PLUG WELDS

A plug weld is a circular weld made either by arc or gas welding through one member of a lap or tee joint. Plug-weld holes in thin plates are completely filled; in heavier plates ($\frac{3}{8}$ in. thick and over) they are only partially filled.

Plug welds in vessel construction are used most often to fix the corrosion resistant lining to the base metal. They are sometimes used as strength welds in single lap joints, reinforcing pads, or nonpressure structural attachments, and then only in addition to other types of welds.

Determining Plug Weld Size

Code Division 1 allows only 30 percent of the total load to be carried by plug welds if they are used (UW-17). The allowable working load on a plug weld in shear or in tension can be computed by the Code formula:

$$P = 0.63(d - \tfrac{1}{4})^2 S_a$$

where

S_a = the allowable stress in tension of the base metal
d = the bottom diameter of the weld hole, limited to $T + \frac{1}{4} < d < 2T + \frac{1}{4}$
T = the plate thickness being welded.

The weld spacing should be equal or larger than $4d$.

Similar to plug welds, but not too often used in pressure vessel design, are slot welds for transmitting larger loads than plug welds. Fillet welds in slots or holes cannot be considered as plug or slot welds.

Example 10.6. Determine the combined load capacity of the plug weld and fillet weld in the single lap joint shown in Fig. 10.14.

Plate materials: SA515 grade 60, S_a = 15,000 psi, E = 0.50 based on weld leg.

Allowable force on plug weld: $P = 0.63(d - \tfrac{1}{4})^2 \, 15,000 = 5,320$ lb.

Allowable force on fillet weld: $F = 4 \times 0.4375 \times 0.50 \times 15,000 = 13,130$ lb.

Total allowable load per spacing: $P + F = 18,450$ lb.

$$(5,320/18,450) \times 100 = 29 \text{ percent} < 30 \text{ percent.}$$

Efficiency of the joint: $(18,450/4 \times \tfrac{1}{2} \times 15,000) \times 100 = 62$ percent.

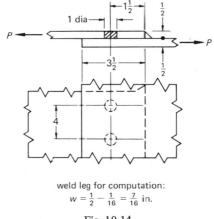

weld leg for computation:
$$w = \tfrac{1}{2} - \tfrac{1}{16} = \tfrac{7}{16} \text{ in.}$$

Fig. 10.14.

10.5. DESIGN ALLOWABLE STRESSES FOR WELDED JOINTS

An important responsibility of the designer is to select a proper allowable work-ing stress or safety factor for the pressure or structural attachment welds used in the vessel design. Weld metal properties should equal or be very close to those of the base metal. However, usually too much attention is given to the higher strength of the weld deposit metal and too little to the heat-affected critical areas of the adjacent metal.

It is advantageous for the designer to be able to express the strength of the weld material in terms of the strength of the base metal modified by some efficiency factor E to compensate for possible variations in the weld quality and approximations in the stress computations.

Code Division 1 specifies in Table UW-12 and paragraph UW-12 the maxi-mum allowable joint efficiencies for welded joints of main pressure vessel seams, with certain limitations. The joint efficiencies depend on the type of weld and the degree of radiographic examination. Paragraph UW-15 gives the allowable stress values in terms of percentages of the plate material for groove and fillet welds of nozzle and other connections. Paragraph UW-18 specifies the allowable load on fillet welds connecting nonpressure parts to vessel pressure parts as equal to the product of the weld area based on the minimum leg dimension w, the allowable stress in tension of the material being welded S_a, and the joint effi-ciency of 55 percent.

Code Division 2 requires for all pressure shell welds a full radiographic exami-nation. The strength of welds so inspected is considered to be the same as the strength of the base metal (AD-140). All welds, groove or fillet, joining non-pressure parts to pressure parts (AD-920) must be continuous, with strength defined as the design stress intensity value of the welded material (weld metal if weaker) times the nominal weld area and multiplied by a Code reduction factor. Nominal weld area for fillet welds is the minimum throat area, with t not less than $\frac{1}{4}$ in. Welds subjected to fluctuating stresses must be designed and evaluated according to design values based on fatigue analysis, as described in Code Appendix 5.

AISC Specifications [34] are often used for the design of structural parts connected to a vessel, such as lifting lugs, supports, etc. The permissible design stress for groove welds with full penetration can be taken the same as for the parent material if prescribed electrodes are used. The allowable stress for the fillet weld throat area is given in terms of the specified strength of the weld metal, depending on the electrode used. Here the nominal composition of the electrode must be considered by the designer when selecting the joint efficiency for the weld.

Summary. Allowable joint efficiencies for welds are usually given in the applicable codes and specifications. In the absence of definite rules the designer has to estimate the efficiency E. A good engineering practice would be to select (in terms of decimal fractions):

for groove welds: $E = 0.80–1.00$

for fillet welds: $E = 0.60–0.80$ (based on throat area)

  $E = 0.45–0.55$ (based on leg area)

for plug welds: $E = 0.60–0.80$

for forged welds: $E = 0.80$.

The upper efficiency values would be used for welds examined by radiographic, magnetic particle, or liquid penetrant examinations.

The term *joint efficiency* is considered by some as a holdover from riveted construction with a definite ultimate strength and is replaced by terms such as *reduction* or *reliability factor*.

The allowable stresses for welds under fluctuating loads will be substantially lower than the allowable stresses for welds subject to static loads only. They will be based on the endurance strength S_N of the base metal and related to the number of working cycles N. For instance, the endurance limit of butt-welded joints could be taken as $S'_N = 0.85S_N$. In addition, concentration factors have to be used in computing the maximum stress in the structural part.

10.6. STRESS CONCENTRATION FACTORS FOR WELDS

Stress concentrations in welds are due to geometry of welded joints, defects and imperfections in the welds, and also the different metallurgical structures of the weld metal, the metal in the heat-affected zone, and the base metal.

Table 10.6. Stress Concentration Factors.

Reinforced full-penetration groove-weld butt joint	$\beta = 1.2$
Toe of transverse fillet weld	$\beta = 1.5$
End of parallel fillet weld	$\beta = 2.7$
Groove-weld T-joint with sharp corners	$\beta = 2.0$

The effects of these factors on stresses induced by steady loads in a ductile weld material can be disregarded. However, if the weld is hard and brittle, or under shock, or fluctuating loads, the influence of stress concentrations becomes significant. The stress concentration factors in Table 10.6 are used to ensure safer welded connections.

10.7. DEFECTS AND NONDESTRUCTIVE EXAMINATIONS OF WELDS

The most common weld defects can be summarized as follows:

poor weld shape due to misalignment of the parts being welded,
cracks in welds or heat-affected zones of the base metal,
pinholes on the weld surface,
slack inclusions or porosity in the form of voids,
incomplete fusion between weld beads or weld and base metal,
lack or insufficient penetration of the weld metal in joints,
undercutting, an intermittent or continuous groove adjacent to the weld left
  unfilled by weld metal.

A good weld design starts with the designer. However, the final well welded connection depends on many factors. Well prescribed welding procedure, good fit-up of parts being welded, supervision, and weld examination are required to ensure good quality welds in addition to good design. The designer usually specifies the type of examination called for by the applicable code or as required for important strength welds. A brief review of the five most important basic methods of inspection follows.

Thorough *visual inspection* (*VT*) is usually satisfactory for minor structural welds. Also, all surfaces of welds to be further examined are first thoroughly visually inspected. The *liquid penetrant examination* (*PT*) is used to examine more important welds or welds where other methods would be hard to apply. Both of these methods are suitable for detecting cracks or any surface defects or subsurface defects with surface openings. If liquid penetrant inspection is used on multilayer welds, each bead should be so examined to inspect the weld properly. In practice at least the root and the finish passes are routinely inspected. In vessel construction a liquid penetrant test is required after a hydrotest.

The *magnetic particle test* (*MT*) is suitable for detecting cracks, porosities, and lack of fusion at or near the surface in ferromagnetic materials. Since this method depends on the magnetic properties of the material tested, it can not be used on nonmagnetic materials such as austenitic stainless steels or aluminum alloys. Again, at least the root and the finish passes of more important welds must be inspected.

The most important nondestructive test method is *radiography (RT)*. It automatically provides a permanent record of the internal quality of the weld material in the form of radiographic films. Any change in density of the weld metal being examined, such as gas pockets, slag inclusions, and incomplete penetrations, will be detected on radiographic film as a dark spot. A good radiographic examination is hard to achieve on tee joints, and therefore radiography is used mainly on butt joints. However, radiographs cannot be interpreted to ascertain the absence of cracks at an angle unless many angle shots are taken.

Ultrasonic examination (UT) is frequently used today to detect subsurface flaws such as laminations or slag inclusions in thick plates, welds, castings, etc.

These nondestructive examinations are not mutually exclusive, but rather complementary.

10.8. WELDING PROCESSES

It would seem appropriate for a vessel designer to have a general knowledge of the main welding processes used in fabrication of pressure vessels. Both the type of welding process used and the welding conditions greatly affect the quality of welds and the estimated weld efficiencies in strength computations of non-Code welds. Selection of proper welding electrodes is usually done by welding engineers. The following discussion presents brief descriptions of the welding processes most commonly used in the fabrication of pressure vessels (UW-27 Welding Processes).

Shielded Metal Arc Welding (SMAW)

This form of welding is widely used. The heat for welding is produced by the resistance of the arc air gap to the flow of electric current. Also called *stick electrode welding*, SMAW is almost always done manually. As the electrode heats, the core wire which conducts the current to the arc melts and provides filler metal for the welded joint. The coating of the electrode breaks down to form a gaseous shield for the arc and weld puddle as well as a small amount of slag, which protects the weld as it cools. Shielding is very important for the quality of weld, since it prevents absorbtion of gases from the air into the molten metal and also prevents the loss of alloying elements during the transfer of molten metal through the arc. The process is very versatile and provides welds of very good quality. However, the rate of weld deposition (the weight of weld metal deposited in a unit welding time) is only moderate, so the use of SMAW is confined to production of small pieces such as nozzles, structural attachments, etc., or to repair work.

Submerged Arc Welding (SAW)

This process, almost always fully automatic, is used in the fabrication of main vessel seams. It gives excellent welds at low cost. However, it can be used for horizontal positions only. A continuous consumable wire coil is used as electrode. Weld puddle and arc are protected by liquid slag, formed from granular mineral flux deposited ahead of the arc. The rate of weld deposition is high.

Gas Metal Arc Welding (GMAW)

A consumable continuous wire is used as an electrode which melts and supplies the filler metal for the welded joint. A protective shield of inert gases (helium, argon, CO_2, or a mixture of gases) is used. The process produces excellent welds at less cost than the GTAW process (see below) with higher weld deposition rate.

Gas Tungsten Arc Welding (GTAW)

This process is used when the highest-quality welding with difficult to weld metals is required. An arc is formed by a nonconsumable tungsten electrode, which carries the electric current; the filler metal, if required, is added separately from a rod or a continuous wire. Inert gas flows around the arc and the weld puddle to protect the hot metal. Weld deposition rate is comparatively low.

Gas Welding

Heat of fusion is generated by burning a flame of gas with oxygen. Different gases are used, as described below.

The oxyhydrogen process (OHW) uses hydrogen for combustion. The highest temperature obtainable is about 4,000°F. OHW is suitable for metals with low melting points, such as aluminum.

The oxyacetylene process (OAW) uses acetylene gas. The maximum obtainable temperature is about 6,300°F, suitable for welding most commercial metals. No flux is used when carbon steel is welded. Almost always used manually for small shop or maintenance welding, and suitable for all positions, OAW requires manual skill. Welds are of good quality, but weld deposition rate is relatively low.

The oxyacetylene flame is also used for *flame cutting* or *flame machining*, which are important processes in the fabrication of steel. Flame cutting is basically a chemical process. Oxygen is fed to the heated metal area through a central orifice in the cutting torch; it oxidizes the heated metal, and the gas pressure forces the oxidized and melted metal out of the cut. Flame cutting, either manual or automated, can achieve high accuracy. When low-carbon steel is flame cut, no detrimental effect in the heat affected zone can be assumed.

Resistance Welding

The heat of fusion is generated by the resistance at the interface to the flow of electric current. No shielding is required. Pressure must be applied for good metal joining. Usually the process is confined to certain jobs and special equipment is provided. *Resistance spot welding* (RSW) or *resistance seam welding* (RSEW) are used to fix corrosion-resistant linings to the wall of a vessel shell.

10.9. WELD SYMBOLS

The standard welding symbols established by the American Welding Society are commonly used by vessel designers to convey the welding information in a concise form on shop or engineering drawings. The reader is referred to refs. 80 and 81 for a listing of these symbols.

11
Selection of Construction Materials

11.1. GENERAL CONSIDERATIONS

This chapter should serve as a brief information guide for a vessel engineer who must be familiar with commonly used construction materials to be able to specify them correctly on engineering drawings or in material specifications for a particular job. A more extensive treatment of the selection of materials for industrial pressure vessels is beyond the scope of this book, see ref. 83 to 100.

The selection of construction materials for Code pressure vessels has to be made from Code approved material specifications. A metallurgical engineer usually specifies the most economical materials of low first cost and/or low future maintenance cost that will be satisfactory under operating conditions and will meet other requirements.

There are many factors supported by experience and laboratory test results that must be considered in selecting the most suitable materials. They include the following:

corrosion resistance in the service corrosive environment,
strength requirements for design temperature and pressure,
cost,
ready market availability,
fabricability,
quality of future maintenance.

Generally, process equipment is designed for a certain minimum service life under specific operating conditions. Based on a corrosion rate in mils (0.001 in.) per year (MPY) a total corrosion allowance is established which is added to the calculated required thickness. Typical design lives are given below for several types of petrochemical equipment:

20 years: Fractionating towers, reactors, high-pressure heat-exchanger shells, and other major equipment which is hard to replace.

10–15 years: Carbon-steel drums, removable reactor parts, and alloy or carbon-steel tower internals.

5-10 years: Carbon-steel piping, heat-exchanger tube bundles, and various process column internals.

The selected material must be suitable for services of different levels of severity from the standpoint of pressure, temperature, corrosive environments, cyclic or steady operations, etc. Obviously, a number of divisions is possible. However, since the choice of material for a vessel depends primarily on the service environment, it would seem practical to classify construction materials according to service: *noncorrosive*, with corrosion rates negligible or very low and definitely established (for carbon steel, a maximum of $\frac{1}{4}$ in. total; otherwise an alternative material with a better corrosion resistance is used); or *corrosive*, requiring special materials other than carbon steels or low-alloy steels.

11.2. NONCORROSIVE SERVICE

In addition to corrosion resistance, the fundamental material selection criteria are design temperature and design pressure.

In the range of *cryogenic temperatures* (from $-425°F$ to $-150°F$) carbon and low alloy steels are brittle and austenitic stainless steels or nonferrous metals like aluminum alloys that do not exhibit loss of the impact strength at very low temperatures must be employed, (see Table 11.1). (For a cyrogenic engineer the dividing line between the cryogenic and low temperatures is usually $-240°F$, below which temperature only so-called permanent gases remain in the gaseous state. This distinction is not of practical significance here.)

The temperature range at which a material changes gradually from ductile to brittle is called the *transition temperature* and is readily determined from Charpy impact tests conducted over a range of temperatures. The designer of Code low-temperature equipment must base his computations on the Code approved properties of the material at room temperature. However, in non-Code work the higher yield and tensile strengths of alloys at very low temperatures can be used to reduce weight and cost where possible. Because of the low reactivity of most chemicals at very low temperatures, corrosion problems are few.

At *low temperatures* (from $-150°F$ to $+32°F$; the Code upper limit is $-20°F$), low-alloy and fine-grain carbon steels tested for notch toughness are found to perform satisfactorily.

In the range of *intermediate temperatures* (from $+33°F$ to about $+800°F$) low-carbon steels are sufficient. Up to about $800°F$ they behave essentially in an elastic manner; that is, the structure returns to its original dimensions when applied forces are removed and maximum stress is below the yield point. The design allowable stress is based on the yield strength or the ultimate strength, obtained from short time rupture tests, supplemented by fatigue or impact tests, where fluctuating or shock stresses are involved.

At *elevated temperatures* (above 800°F) marked changes in mechanical properties occur in steels. They begin to exhibit a drop in ultimate and yield strengths and cease to be elastic, becoming partly plastic. Under a constant load, there is a continuous increase in permanent deformation, called *creep*. The *creep rate* is measured in percent of a unit length per unit time. Actually, some creep begins at temperatures over 650°F, but it does not become an important factor for carbon steels until temperatures over 800°F are reached. The design allowable stress is then based on two criteria: (a) the deformation due to creep during the service lifetime must remain within permissible limits, and (b) a rupture must not occur. The allowable stresses are obtained from long-term creep tests and from stress rupture tests at elevated temperatures. Few data, if any, are available on high-temperature endurance limits.

Selection of steels for elevated-temperature service is generally a complex problem and not so straightforward as material selection for lower temperatures. The choice has to be based on several factors. At high temperatures, a number of changes in the steel microstructure may occur which affect mechanical properties to a high degree. The mechanical properties of alloys are affected both by chemical composition and by grain size. Usually, at low and intermediate operating temperatures, fine-grained microstructure (above ASTM No. 5) is preferred in steels because of the resulting higher tensile, fatigue, and impact strengths and better corrosion resistance. However, at high temperatures, where the main requirement is a superior creep–rupture strength, a coarsed-grained material may be preferred.

Steels used in vessel construction for elevated temperatures can be classified into five general types:

1. Carbon steels. These vary in strength at temperatures below 650°F because of small differences in carbon content, but they all have similar properties in the creep range. Where their use is not limited by sulfur corrosion or hydrogen attack, they usually represent the most economical material for intermediate as well as for elevated temperatures at low pressures. Not only are they relatively cheap per pound, they are also comparatively easy to fabricate. Each additional alloying element increases the cost of the steel, and often the difficulty of fabrication and welding as well. The final overall cost of a carbon steel vessel may be much less than the cost of an alloy steel vessel.

2. Carbon–molybdenum steels, low chromium–molybdenum alloy steels (up to 3Cr–1Mo) and *intermediate chromium–molybdenum alloy steels* (up to 9Cr–1Mo). Some of these can be used up to 1200°F, where resistance to graphitization and hydrogen attack is required. These steels have better creep–rupture properties and high-temperature strength than carbon steels, and there is an economy in using them for pressure vessels subjected to high pressure at temperatures over 650°F. Furthermore, these steels may be required to resist oxidation, sulfidation, or hydrogen attack.

Table 11.1. Construction Steels for Noncorrosive Service.

	SERVICE TEMPERATURE (°F)	PLATE	PIPE	FORGINGS	PRESSURE BOLTING	STRUCTURAL BOLTS	STRUCTURAL NUTS	STRUCTURAL SHAPES
Cryogenic	−425 − −321	SA240 types 304, 304L 347	SA312 types 304, 304L, 347	SA182 grades F304, F304L, F347	Bolts: SA320 gr. B8 strain hardened	Same as pressure parts		
Cryogenic	−320 − −151	SA240 types 304, 304L, 316, 316L SA353	SA312 types 304, 304L, 316, 316L	SA182 grades F304, F304L, F316,	Nuts: SA194 gr. 8 (S5 SA20)	Same as pressure parts		
Low temperatures	−150 − −76	SA203 gr. D or E	SA333 gr. 3	SA350 gr. LF3	Bolts: SA320 gr. L7 Nuts: SA194 gr. 4			
Low temperatures	−75 − −51	SA203 gr. A or B	SA333 gr. 3	SA350 gr. LF3				
Low temperatures	−50 − −21	SA516 all grades impact tested (see note 2)	SA333 gr. 1	SA350 gr. LF1 or LF2				
Low temperatures	−20 − +4	SA516 all grades over 1 in. thick impact tested		SA350 gr. LF1 or LF2				
Low temperatures	+5 − +32	SA516 all grades over 2 in. thiok impact tested						
Intermediate	+33 − +60	SA285 gr. C, $\frac{3}{4}$-in. thick max. SA515 gr. 55, 60, 65, 1.5-in. thick max. SA516 all grades, all thicknesses	SA53 (seamless) or SA106	SA181 gr. I or II SA105 gr. I or II		SA307 gr. B. or SA325		SA36
Intermediate	+61 − +775	SA285 gr. C, $\frac{3}{4}$-in. thick max. SA515 gr. 55, 60, 65, 1.5-in. thick max.	SA53 (seamless) SA106 SA335 P1	SA181 gr. I or II SA105 gr. I or II	Bolts: SA193 gr. B7 Nuts: SA194 gr. 2H	SA307 gr. B. or SA325		SA36

Elevated temperatures					
+776–+875	SA516 all grades, all thicknesses; SA204 gr. B, all thicknesses		SA182 gr. F1		Same as pressure parts
+876–+1000	SA204 gr. B or C	SA335 P1	SA182 gr. F11 / SA182 gr. F12	SA193 gr. B5 / SA194 gr. 3	
+1000–+1100	SA387 gr. 11 Cl.1 / SA387 gr. 12 Cl.1	SA335 P11 / SA335 P12	SA182 gr. F22	SA193 gr. B8 / SA194 gr. 8	
+1100–+1500	SA240 types 304, 316, 321, 347; 347 preferred (see section 11.5)	SA312 types 304H, 316H, 321H, 347H	SA182 grades 304H, 316H, 321H, 347H		
above +1500	Type 310 stainless; Incoloy				

1. Stainless steels types 304, 304L, and 347 and 36 percent Ni steels for service temperatures below −425° F must be impact tested.
2. Pressure-vessel steel plates are purchased to the requirements of the standard ASTM (SA) A20, which requires testing of individual plates. For low-temperature service, carbon steel material is ordered to meet the impact requirements of supplement S of the standard ASTM (SA) A20. Typical material specification is as follows: "SA516 gr. 60, normalized to meet impact requirements per supplement (S5) of SA20 at −50° F."
3. The limiting design temperature is determined by the behavior of the metal in the particular environment and its corresponding mechanical properties. High-temperature limitations are hard to define accurately. For instance, to avoid graphitization or hydrogen-attack problems some metallurgists recommend 550° F as maximum design temperature for carbon-steel parts in petrochemical plants.
4. Materials for structural attachments welded to pressure parts and transmitting loads during operation are generally of the same grade as the pressure parts. All permanent attachments welded directly to 9 percent nickel steels should be of the same material or of an austenitic stainless steel type which cannot be hardened by heat treatment.
5. Material SA36, if used at lower temperatures (below 32° F), should be of silicon-killed fine grain practice.

3. Ferritic (straight chromium) stainless steels. These are used in some applications.

4. Austenitic stainless steels. These are the only steels assigned allowable stresses in the Code for temperatures higher than 1200°F, up to 1500°F. A decrease in oxidation resistance limits their usefulness above this temperature.

5. Special high-temperature-resisting alloys. These are used for temperatures above 1500°F. They include type 310 stainless steels and Incoloy.

A large number of high-temperature vessels are fabricated of cheaper, low-alloy carbon steels, for instance SA 204 gr. B, with an internal refractory lining thick

Table 11.2. Description of Plate Materials.

ASTM (ASME) SPECIFICATION	TYPE OF STEEL	REMARKS
A(SA)285 gr. C	Carbon	General pressure vessel material; max. available thickness 2 in; 0.28 percent C max.
A(SA)515 gr. 55, 60, 65, 70	Carbon	Coarse austenitic grain size (ASTM No. up to 5)
A(SA)516 gr. 55, 60, 65, 70	Carbon	Fine austenitic grain size (ASTM No. above 5)
A(SA)283 gr. C or D	Carbon	Structural, supports, clips
A(SA)36	Carbon	Structural steel, various shapes; 0.26 percent C max.
A(SA)204 gr. A, B, C	$C-\frac{1}{2}Mo$	Elevated-temperature service
A(SA)302 gr. B	$1\frac{1}{4}Mn-\frac{1}{2}Mo$	Steam boilers
A(SA)387 gr. 2	$\frac{1}{2}Cr-\frac{1}{2}Mo$	Max. 1000°F Elevated-temperature service.
gr. 12	$1Cr-\frac{1}{2}Mo$	1200°F Each grade is available in two
gr. 11	$1\frac{1}{4}Cr-\frac{1}{2}Mo-Si$	1200°F classes, Cl.1 and 2, of different
gr. 22	$2\frac{1}{4}Cr-1Mo$	1200°F tensile strength, depending on
gr. 21	$3Cr-1Mo$	1200°F the heat treatment, Cl.2 has a
gr. 5	$5Cr-\frac{1}{2}Mo$	1200°F course austenitic grain size
A(SA)442 gr. 55 or 60	Carbon	Fine grain practice (FGP) quality preferred for low-temperature service
A(SA)203 gr. A or B	$2\frac{1}{2}Ni$	Min. operating temperature −90°F
A(SA)203 gr. C or D	$3\frac{1}{2}Ni$	−150°F
A(SA)353	9Ni	−320°F
A(SA)410	CrCuNiAl	Max. 0.12 percent C, min. operating temperature −125°F

enough to keep the wall metal temperature at a level where the full strength of the metal can be used in computing the shell thickness. Internal insulation is not practical, however, in heat exchangers or small-diameter piping, where stainless steels are more often used. Where an internal refractory is used, differential thermal expansion of internals must be considered, as well as heat transfer through internals welded to the shell, which cause shell hot spots.

Table 11.1 summarized the steels most commonly used in the vessel design at various temperatures. Table 11.2 provides a guide for quick identification of steels by type.

11.3. CORROSIVE SERVICE

Glass, rubber, enamel, lead, and teflon are successfully used as protective lining materials. The use of such linings calls for specialized methods of fabrication. However, the most common and commercially readily available corrosion-resistant materials for petrochemical plants are stainless steels. For up to $\frac{3}{8}$-in.-thick vessel shells it is most economical to use solid stainless steel plate; above this thickness carbon or low-alloy steel shells with applied corrosion-resistant layers are used.

In general, three methods of attaching the protective layer to the carbon steel plate are usually applied; integral cladding, strip or sheet lining, and weld overlay cladding.

Integrally Applied Cladding

Integrally clad or rollclad plate is fabricated in a steel mill by hot rolling of assemblies of carbon or low alloy steel plates (backing) and corrosion resistant sheet (linear) which have been welded at the edges. In rolling at high temperatures the pressure creates a solid-phase weld between the backing and cladding metals. The bond strength usually exceeds the minimum 20,000 psi in shear strength required by the Code. The thickness of the liner can be specified in percent of backing-plate thickness, or more often in fractions of an inch. For most applications a thickness of $\frac{5}{64} - \frac{1}{8}$ in. is quite sufficient and is considered a corrosion allowance. Rarely does the thickness of an integrally applied liner exceed $\frac{3}{8}$ in.

Integrally clad plate has many advantages. It can be hot or cold formed; lightly loaded internal support clips can be directly attached to the cladding; and the continuous bond eliminates the possibility of any corrosive substance getting between the cladding and the backing metal. It would seem that differences in thermal expansion might introduce peak stresses leading to cracks in the cladding, particularly in the region of welds, but based on practical experience, the integral cladding performs quite well at the elevated temperature or cyclic service. However, special tests may be prescribed to prove that integrally clad plate is suitable for the intended operating corrosive conditions. Also, frequent service inspections of the cladding are recommended. Diffusion of carbon from ferritic backing steels into austenitic stainless cladding occurs at high bonding temperature, but the carburized zone is very narrow and does not affect the overall strength or the corrosion resistance of the clad plate. Carbon diffusion can be minimized by nickel plating of the backing plate.

Where mild corrosion resistance is required, most cladding materials are made of straight chromium steels, types 405 or 410S. Both have lower thermal expansion than carbon steel. For more severe corrosive service 18-8 stainless steels are used. The grades with low carbon content or the stabilized grades have to be used if welding is used in fabrication or a post weld heat treatment is required. PWHT temperatures may be in the carbide precipitation range for unstabilized austenitic stainless steel and/or in the range of sigma-phase formation; proper attention has to be given to this in the selection of the stainless steel cladding material, otherwise heat treatment may result in inferior mechanical and corrosion properties of the cladding material.

Table 11.3 lists some frequently used cladding materials; the backing material can be any carbon or low-alloy steel.

Strip or Sheet Lining

An alternative method to provide pressure vessels with a corrosion-resistant layer is a strip-type or sheet-type lining attached by welding to the vessel shell. The thickness of the alloy sheet or strip is usually $\frac{5}{64} - \frac{1}{8}$-in. thick. Any liner material used for integrally clad plates can be used for sheet or strip lining.

Strip-Type Lining. Strips 3-5 ft long by 3-6 in. wide, depending on service temperature (narrower strips for higher temperatures), are attached to the vessel wall by a continuous weld around the edges. The welds between the strips are $\frac{1}{4} - \frac{1}{2}$ in. wide. Filler weld metal should be of the same chemical composition as the liner. A soap-suds air test is conducted after completion of the weld to reveal possible cracks and leaks in the liner or backing plate.

Sheet-Type Lining. The sheets are several feet in width and length, tightly fitted, and attached to the vessel wall by resistance spot welds on $1\frac{1}{2}$ in. $\times$ $1\frac{1}{2}$ in. square spacing for temperatures up to 800°F and $1\frac{1}{2}$ in $\times$ 1 in. spacing for temperatures above 800°F. Sometimes $\frac{1}{2}$-in. plug welds or seam resistance welds are used. The purpose of spot welds in addition to the circumferential attachment welds is threefold:

1. The spot welds provide a tight attachment against the vessel wall.
2. Usually the protective sheet lining material has a different thermal expansion rate than the carbon steel backing and the alloy liner would buckle

Table 11.3.

ASTM (SA) STANDARD SPECIFICATIONS FOR INTEGRALLY CLAD PLATE	ALLOY CLADDING METAL ASTM (SA)	REMARKS	MAXIMUM SERVICE TEMPERATURE (°F)
Chromium steels,	A (SA) Type 410S	0.08 C max.	1200
A (SA) 263	405	12Cr–Al	1200, note 4
	429	15Cr	notes 3, 5
	430	17Cr	notes 3, 5
Chromium-nickel	A (SA) 240 Type 304L	0.03 C max.	800
steels, A (SA) 264	309S	23Cr–12Ni	note 6
	310S	25Cr–20Ni	note 6
	316L	0.03 C max.	800
	321	Ti stabilized	note 6
	347	Cb stabilized	note 6
Nickel and nickel-base	B (SB) 127	Monel 400	500
alloys, A (SA) 265	168	Inconel 600	1200
	409	Incoloy 800	1200
	424	Incoloy 825	
Copper and copper-base	B152 Cu alloy	P-deoxidized	
alloys, B432	No. 122	copper,	
	No. 102	oxygen-free,	
	B402 No. 706	90Cu–10Ni	
	No. 715	70Cu–30Ni	

1. Internal parts (trays, pans, baffles, etc.) are of the same class of alloy as the shell lining.
2. Maximim service temperature of the lining can be controlled by the maximum service temperature allowed for the backing material.
3. The Code does not recommend the use of the Cr-stainless steels with a content of Cr over 14 percent for service metal temperature over 800°F.
4. Operating temperatures should be outside the temperature range from 750°F to 950°F, where the material is subject to embrittlement.
5. Not suitable for welding.
6. Because of the different thermal expansions of austenitic steels and carbon steels and graphitization of carbon steels the operating temperature is also quite often limited to 800°F.

under high operating temperature. The spot welds keep the liner from buckling and force the liner sections to compress elastically; in time this changes into partly plastic deformation.

3. The spacing of the spot welds should be close enough to protect the liner against permanent buckling in case of seepage of the test liquid behind the liner and a sudden release of the pressure of the liquid in the vessel during a hydrotest.

Usually, sheets and strips are welded to the vessel wall after the vessel is entirely welded and otherwise completed. However, sheets can be attached to the base plate before rolling or forming. Carbon-steel surfaces are prepared (by grinding) to provide a suitable surface for application of the liner. A vessel fabricated in this way is less expensive than a vessel constructed from integrally clad plate; however, where complete tightness is required integrally clad plate construction is preferred.

Weld Overlay Cladding

Weld overlay cladding is another frequently used method to produce a continuous bonded layer of corrosion- or wear-resistant alloy on a base metal. In the construction of pressure vessels this method is particularly used as a supplement to the previous methods; for instance, where the liner has been stripped to allow a support bracket to be welded directly to the base metal, it must be replaced by a weld deposit of the same chemical composition. With a weld overlay irregular shapes can be covered without difficulty.

There are several potential disadvantages to weld overlay cladding:

1. The procedure may distort and introduce large residual internal stresses into the clad object.

2. A metallurgical reaction may take place, such as carbon diffusion from the ferritic base to the austenitic weld, with resulting pollution of the alloy layer. The thickness of the weld overlay is usually $\frac{3}{16}-\frac{1}{4}$ in. thick to provide a minimum $\frac{1}{8}$ in. of chemically pure cladding.

3. Differences in coefficients of thermal expansion become important for high-temperature service (over $800°F$). Each time the austenitic weld overlay on carbon steel is heated and subsequently cooled, shear stresses of yield magnitude arise at the fusion line.

Nozzles of small size ($\leqslant 4$ in.) are usually made of solid alloy steel up to $800°F$ design temperature; larger nozzles are lined with weld-overlayed flanges.

The weld overlay method is the most expensive of the three methods described.

11.4. BOLTING MATERIALS

Bolts for Pressure Connections

Bolting material for pressure connections must conform to the specifications listed in the Code. Bolts are designed not only for strength but also for tightness at the joints. In order to prevent leakage of a bolted joint, the total force exerted by the bolts must exceed the sum of the force due to the operating fluid pressure and the force necessary to keep the joint tight. The latter depends on the gasket material and the design of the joint.

Bolts have to be tightened to some initial elastic strain and corresponding elastic stress. In high-temperature service, or over a long period of time, creep occurs in the bolts and some of the elastic strain is transformed into a plastic strain, with a corresponding reduction in stress and tightening force. This is called *relaxation*. When relaxation occurs to the extent that the total force cannot keep the joint tight, the joint leaks. Therefore, a high resistance to relaxation is important. Since the relaxation can occur also in the flange and gasket their properties should be checked as well.

Other important properties of bolting materials to consider in the design are yield strength, notch toughness, notch sensitivity, ductility, and coefficient of thermal expansion.

Allowable bolt stresses permitted by the Code include high safety factors and take into account the fact that additional torsion and bending stresses induced in bolts during the pretightening may become severe and also that an additional retightening will be required. A maximum design temperature of 450°F was specified for ordinary carbon steel bolting material because of its low resistance to relaxation.

Austenitic stainless steels have low yield strength and tend to gall at higher temperatures. Generally, they are used on pressure joints for low temperatures or for corrosive service. For best results, the preload stress on high-temperature bolting should be controlled. For other temperatures, handwrenching is usually satisfactory. Since galling is a problem with stainless steels (particularly 300 series) it is a common practice to use slightly different alloys for bolts and nuts. It should be remembered that unlike metals are more satisfactory in sliding contact under pressure than are like metals.

In practice, SA193 gr. B8 (type 304) bolts are used with SA194 gr. 8F (type 303) nuts and their compositions are close enough to cause galling. To minimize galling when the nuts are tightened, stainless steel fasteners are made up with a thread lubricant such as Molykote Type G-n Paste.

Tables 11.4 and 11.5 describe most often used bolt and nut materials.

As previously mentioned, as operating temperature rises either leakage occurs or additional stress in bolts, gasket, and flange is set up when materials with different coefficients of thermal expansion are used for bolting and flanges. Specification A453 with four grades (660, 651, 662, 665) covers a bolting material with thermal expansion coefficient comparable to austenitic stainless steels (300 series). This material, a steel with high Ni and Cr content, requires special processing, and is not intended for general-purpose applications, but for important joints where the additional safety factor is justified. For instance, it is used by some valve manufacturers for bolting for nuclear austenitic stainless steel valves. Specification A437 with two grades (B4B and B4C), a high-alloy bolting material with 12 percent Cr could be used for special high-temperature, high-pressure service with intermediate chrome-steel constructions of the same heat expansion properties.

Table 11.4.   Bolts for Pressure Joints.

ASTM (SA) SPECIFICATION	TYPE OF STEEL	ALLOWABLE SERVICE METAL TEMPERATURE (°F)	REMARKS
A(SA) 193 gr. B7	1Cr–$\frac{1}{5}$Mo	–20–1000	
B16	1Cr–$\frac{1}{2}$Mo	–20–1100	see note 1;
B5	5Cr–$\frac{1}{2}$Mo	–20–1200	ferritic steels
B6	13Cr (type 410)	–20– 900	
A(SA) 193 gr. B8 Cl.1	Type 304		
B8M Cl.1	316		austenitic steels;
B8T Cl.1	321	–20–1500	see note 2
B8C Cl.1	347		
A(SA) 307 gr. B	carbon	–20– 450	
A(SA) 320 gr. L7	Cr–Mo QT	–150 min.	
B8	Type 304	–425 min.	
B8C	347 sol.	–425 min.	max. service
B8F	303 treat.	–325 min.	temperature
B8M	316	–325 min.	+100°
B8T	321	–325 min.	

Notes:
1. For use at temperatures from –20°F to –50°F, this material must be quenched and tempered.
2. The identification Cl.1 describes solution-treated material. When increased mechanical properties are desired, austenitic bolting can be both solution treated and strain hardened (Cl.2).

Table 11.5.   Nuts for Pressure Joints.

ASTM (ASME) SPECIFICATION (SEE NOTE 2)	TYPE OF STEEL	ALLOWABLE SERVICE METAL TEMPERATURE (°F)
A(SA) 194 gr. 1 or 2	carbon	–20– 900
gr. 2H	carbon, QT	–50–1100
gr. 3	5Cr, QT	–20–1100
gr. 4	C–Mo, QT	–150–1100
gr. 6	12Cr (410), QT	–20– 800
gr. 7	1Cr–$\frac{1}{5}$Mo	–20–1000
gr. 8F	Type 303, S.Tr.	–325– 800
gr. 8M	316, S.Tr.	–325–1500
gr. 8T	321, S.Tr.	–325–1500
gr. 8C	347, S.Tr.	–425–1500
gr. 8	304, S.Tr.	–425–1500

Notes
1. Impact test is required for Type 303 in low-temperature service if the carbon content is above 0.10 percent.
2. These are the product specifications; no design stresses are required.

Bolts for Structural Connections

Bolting for structural nonpressure parts is designed for strength only, since the tightness of the joint is not important. Consequently, less expensive materials can be selected with properties suitable for the design environment. For non-corrosive service, carbon-steel bolting SA307 gr. A or B or SA325, for corrosive or high-temperature service bolting material A193 gr. B5, B6, B8C, B8T, B8 are commonly used.

Internal bolting for lined vessels is of the same material as the lining and the bolting for internals (trays) is usually also of the same material as the internals. The allowable tensile or shear stresses used in stress computations are higher than Code allowable stresses. A good guide for allowable bolting stresses for nonpressure parts can be found in ref. 9.

11.5. STAINLESS STEELS

General Considerations

Steels with a chromium content of 11 percent or more, but less than 30 percent, are known as stainless steels because of their excellent resistance to corrosion. With over 30 percent chromium content they are classified as heat-resisting alloys.

Stainless steels are basically alloys of chromium and iron. The most important additional alloying element is nickel. Other alloying elements may be added, including carbon, manganese, molybdenum, columbium, titanium, selenium, silicon, and sulfur, all of which result in properties required for special service. Stainless steels are frequently used for construction of petrochemical processing equipment and in many other applications. Some of the reasons for this use are: to provide the necessary resistance to a corrosive environment, thus increasing the service life of the equipment and the safety of the working personnel; to provide strength and oxidizing resistance at elevated temperatures and impact strength at low temperatures; to facilitate the cleaning of the equipment.

Stainless steels become corrosion resistant (passive) because of the formation of an unreactive film which adheres tightly to the surface of the metal. This can be chromium oxide or an adsorbed oxygen film that acts as a barrier protecting the metal against further attack in certain types of environment. This protective film, which is not visible, is formed within minutes or months, depending on the type of alloy. The formation of the film may be hastened or it may be produced artificially by a strong oxidizing agent such as solution of nitric acid in water. This artificial formation of protective film, or *passivation*, serves a double purpose, since it also helps to remove any foreign metals or other substances that might contaminate the stainless-steel surface.

If the chromium content is less than 11 percent the film is discontinuous, and the corrosion resistance of such steels approaches the relatively poor corrosion resistance of ordinary steel.

The variation in corrosion resistance appears to be dependent on the amount of the chromium in the steel; it can be greatly improved by the addition of nickel and molybdenum as well. Stainless steel grades 410 or 405 are less corrosion resistant than those having a higher amount of alloying elements, such as the grades 304 or 316.

No protective coatings such as paints are applied to the surface of stainless steel parts, since they would only prevent oxygen penetration for formation of the passive layer. All grades of stainless steel are affected in some manner by welding heat, which modifies their resistance to corrosion and changes their mechanical properties. Welding metal deposits are of cast structure.

Stainless steels are produced mostly by the electric furnace process. The cost of stainless steels varies with type, form, and quantity, and not all the grades are readily available in every form.

Based on the principal alloying constituents the stainless steel used for pressure vessel and piping construction can be divided into three main groups:

1. the straight chromium group (400 series) with chromium content up to 30 percent;
2. the chromium–nickel group (300 series), quite often referred to as 18-8 stainless steels, with Cr and Ni varying in percentage; and
3. the chromium–nickel–manganese group (200 series), with manganese replacing a portion of the nickel.

Based on their metallurgical microstructure, these stainless steels can be classified as austenitic, ferritic, or martensitic.

Austenitic Stainless Steels

Because of their large percentage of nickel, 300 series stainless steels retain their austenitic structure after cooling, with Cr, Ni, and C in solid solution with iron. Under a microscope only austenitic crystals can be distinguished. These are high chromium–nickel–iron alloys. They are nonmagnetic, highly corrosion resistant even at temperatures up to 1500°F, and hardenable only by cold working, and they possess high impact strength at low temperatures. The typical and most commonly used grades of stainless steel are grades 304 and 316. Higher-chromium-content austenitic stainless steels (grades 309 and 310) are resistant to oxidation and sulfur attack up to 2000°F.

The primary problem with austenitic stainless steels is grain-boundary sensitization. Most austenitic stainless steels are furnished by producers in solution-annealed condition. Solution annealing of type 300 stainless steels consists of raising the metal temperature above 1850°F, where austenite acts as a powerful solvent. With Cr, Ni, and C dissolved in the austenitic matrix, these steels offer maximum corrosion resistance together with the maximum ductility and strength. To retain this microstructure at lower temperatures, these stainless steels have to be cooled rapidly to below 800°F. However, at any subsequent rise in the temperature (for instance at welding) to the range of 800–1600°F carbon molecules diffuse to the grain boundaries and precipitate out of the solid solution as chromium carbide (Cr_4C) at the boundaries (see Fig. 11.1). The effect is depletion of chromium content in the thin envelope surrounding each grain. The carbide formed is not as corrosion resistant as the metal from which it develops, and the corrosion resistance of the envelope depleted of chromium is drastically reduced. The stainless steel becomes susceptible to intergranular corrosion and is said to be *sensitized*. If the thin envelope corrodes, the grain rich in chromium falls out and metal begins to disintegrate. The boundary envelope poor in chromium is also anodic with respect to the rest of the grain, and galvanic corrosion is possible.

Sensitization of all the material may be caused by slow cooling from annealing or stress-relieving temperatures. For instance, stainless steel parts welded to a carbon-steel vessel shell can be sensitized by stress relief given to the carbon-steel shell. Welding will result in sensitization of a band of material $\frac{1}{8}-\frac{1}{4}$ in. wide slightly removed from and parallel to the weld on each side (Fig. 11.2). These two areas are the heat-affected zones where the steel has been held in the sensitizing range longer than elsewhere and cooled slowly. The material in between, including the weld metal, is not sensitized, since its temperature is raised well above 1600°F and subsequent cooling is comparatively rapid.

Sensitization may not be harmful in certain environments, for instance if con-

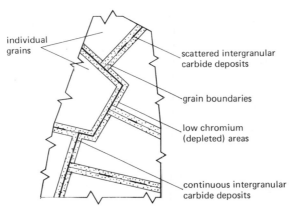

individual grains

scattered intergranular carbide deposits

grain boundaries

low chromium (depleted) areas

continuous intergranular carbide deposits

Fig. 11.1. Schematic representation of the grain structure in Type 300 sensitized stainless steel.

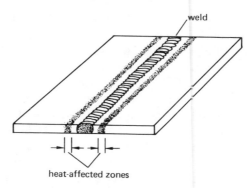

Fig. 11.2. Heat-affected zones, susceptible to intergranular corrosion in austenitic stainless steels.

tinuous exposure to liquids is not involved and when operating temperature does not exceed 120°F.

The corrosion properties of sensitized steel can be restored by *desensitization*, that is, heating above 1600°F to dissolve carbides and subsequent rapid cooling (see Fig. 11.3). The effect of sensitization on mechanical properties is far less important, being almost negligible at intermediate temperatures, and causing some ductility loss at low temperature.

According to the degree of possible sensitization of the grain boundaries, the austenitic stainless steels can be divided into three groups:

Group I. These are the normal-composition, so-called 18-8, chromium-nickel steels, such as typical grades 304, 316, 309, and 310. They are susceptible to sensitization, which means that their corrosion resistance in environments usually encountered in petrochemical plants is reduced by welding or by flame cutting, whether used for preparation of edges that are to be welded or for cutting of openings. To regain full resistance to corrosion, it may be necessary to give the weldment a final full solution annealing. However, the required quick quenching may introduce residual stresses which are too harmful for certain applications. To avoid impairing corrosion resistance, low-temperature stress relieving (below 800°F), holding at that temperature for a relatively long time, and then allowing the weldment to cool slowly, is sometimes used. Obviously, this procedure is not very effective, since the maximum locked-in stresses after a stress relief are equal to the depressed yield strength at the stress-relieving temperature. In comparison with carbon steels, the stainless steels require a much higher stress-relieving temperature and a longer holding time, since they retain their strength at elevated temperatures.

To summarize, the standard 18-8 stainless steels in the solution-annealed state are suitable for parts in corrosive environments, when no welding or stress relief are required and the operating temperatures stay below 800°F.

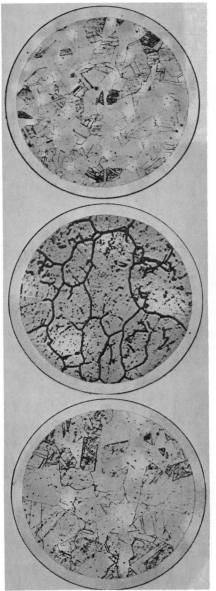

A. Austenite grain structure in fully annealed Type 304 (prior to welding). Outlines of crystal grains as made visible under the microscope by etching prepared specimens with a suitable reagent. This microphotograph shows absence of intergranular carbides.

B. Same as A above, except for heavy black lines surrounding the crystals. These are the locations of carbide deposits along the grain boundaries. Their intensity varies from light, scattered precipitates (dots under the microscope) to heavy lines, as shown, completely surrounding the individual grains. Whether they are scattered or non-continuous or whether they are continuous depends on the carbon content, the ratio of combined chromium and nickel to carbon percent, and the length of time the metal is held within the carbide precipitation range (as explained in the text). The condition shown may be regarded as an extremely severe sensitized condition in Type 304.

C. Same as B above, after the same area has been subjected to a subsequent anneal for the re-dissolving of precipitated carbides and the restoration of corrosion resistance properties. This treatment results in putting the metal back in the condition shown in A.

Fig. 11.3.   Grain structure of Type 304 stainless steel [88]. (Courtesy of Allegheny Ludlum Steel Corporation.)

Group II.   These are the stabilized stainless steels, Types 321 or 347. Grain-boundary sensitization is eliminated by using alloying elements like titanium or columbium which stabilize the stainless steel by preempting the carbon: because of their stronger affinity to carbon, they form carbides in preference to the chromium, which stays in solid solution in iron. The carbides formed do not tend to precipitate at the grain boundaries, but rather remain dispersed through the metal. The creep strength of stabilized stainless steels is superior to that of unstabilized steels. Cb is stronger stabilizing agent than Ti, making Type 347 superior to Type 321.

To summarize, stabilized grades of stainless steel in the annealed condition are immune to intergranular corrosion. They can be welded and stress relieved and cooled slowly in air. They can be annealed locally without sensitization of the adjacent areas. However, under certain *special* heat treating conditions they can be sensitized and become susceptible to a corrosion known as *knifeline attack*. They present some problems when welded, being susceptible to cracking. Their cost is quite high, and therefore they are used only for special jobs, such as for operating temperatures above 800°F. They also tend to lose their immunity to intergranular corrosion when their surfaces are carburized by the process environment.

Group III.   These are extra-low-carbon grades like 304L or 316L. Grain-boundary sensitization can be minimized by using low-carbon stainless steels with 0.03 percent C maximum, at the expense of lowered strength. The rate of chromium carbide precipitation is so retarded that they can be held within the 800–1500°F range for up to several hours without damage to their corrosion resistance.

To summarize, extra-low-carbon stainless steels can be stress relieved, welded, and slowly cooled without significantly increasing their susceptibility to intergranular attack. They are very often used in pressure vessel construction, either as solid plate or for internal lining material. They are more expensive than normal-composition stainless steels because of the difficulty and cost of removing the carbon. However, they are not equivalent to group II, since they are subject to sensitization if the operating temperature remains in the 800–1500°F range for a prolonged period of time. Consequently, the extra-low-carbon grades can be used for applications at operating temperatures up to 800°F.

Ferritic Stainless Steels

Ferritic stainless steels usually include straight chromium stainless steels with 16–30 percent chromium. They are nonhardenable by heat treatment. A typical stainless steel of this group is type 430. The grade quite often used for corrosion-resistant cladding or lining is type 405, which contains only 12 percent chromium;

however, addition of aluminum renders it ferritic and nonhardenable. When type 405 cools from high welding temperatures there is no general transformation from austenite to martensite and it does not harden in air. However, it may become brittle in heat-affected zones because of rapid grain growth. Ferritic steels may become notch sensitive in heat-affected weld zones, and they are also susceptible to intergranular corrosion. Ferritic stainless steels are sensitized by heating to a temperature of 1700°F and then air cooled at normal rates. If they are cooled slowly (in a furnace) their resistance to intergranular corrosion is preserved. Annealing of a sensitized ferritic stainless steel at 1450°F allows chromium to diffuse into depleted parts to restore the corrosion resistance.

Welding of ferritic stainless steels sensitizes the weld deposit and the immediately adjacent narrow bands of base material on both sides of the weld, as shown in Fig. 11.4. The composition of electrodes used for welding ferritic stainless steels is often such as to produce austenitic or air-nonhardening high-alloy weld metal.

Sensitized ferritic stainless steel is much less corrosion resistant then sensitized austenitic stainless steel. The methods used to suppress sensitization in austenitic stainless steels are not effective with ferritic stainless steels. When ferritic stainless steels are heated into the 750–900°F range for a prolonged period of time, notch toughness is reduced. This has been termed 885°F *embrittlement* and has been ascribed to the precipitation of a chromium rich α-prime phase.

Ferritic stainless steels also exhibit lower ductility at low temperatures, which limits their use in the low temperature range. In general, ferritic stainless steels are seldom used in vessel construction, except for corrosion resistant lining or cladding (grades 405 or 410S), heat-exchanger tubing, and vessel internal hardware (trays) for less corrosive environments, since they are not as expensive as austenitic stainless steels. They are magnetic and finished parts can be checked by a magnet.

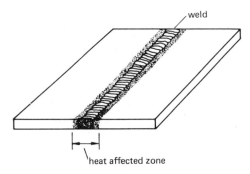

Fig. 11.4.   Heat-affected zone in a straight chromium ferritic stainless steel. The sensitized zone extends across the weld deposit.

Martensitic Stainless Steels

Martensitic stainless steels include straight chromium steels, usually with 11 to 16 percent chromium as alloying element. They are hardenable by heat treatment, that is, their strength and hardness can be increased at the expense of ductility. Type 410 is typical of this group. In the annealed condition at room temperature it has ferritic structure. When heated from 1500°F to 1850°F its microstructure changes to austenitic. If the steel is then cooled suddenly, for instance as in deposited weld metal with adjacent base metal zones in air, part of the austenite changes into martensite, a hard and brittle material. If the cooling is very rapid from 1850°F, the final martensitic content will be at a maximum. Post-weld heat treatment with controlled cooling will reduce residual stresses and will allow the austenite to transform to ductile ferrite. With normal carbon content, the hardenability of straight chromium stainless steels is markedly reduced with above 14 percent chromium. With increased carbon content, they remain hardenable above 14 percent up to 18 percent chromium. With 18 percent chromium content they become non-hardening and their microstructure remains ferritic at all temperatures. The division line between hardenable (martensitic) and nonhardenable (ferritic) steels is not always distinct. Martensitic stainless steels are only rarely used as construction material for pressure vessel parts. They are the least corrosion resistant of the stainless steel grades, and if welding is used in fabrication heat treatment is required. They are most corrosion resistant when in hardened and polished condition. They are magnetic and easily checked by a magnet.

Table 11.6 lists for the nominal composition of individual stainless steel grades as well their approximate price ratios, based on the price of plate.

Alloy Selection, Design, and Fabrication

Most failures of stainless steels can be traced to the wrong selection of material, to poor design, or to improper fabrication and handling.

Selection of the right grade, with possible heat treatment for a particular service, is usually done by a metallurgical engineer or a process engineer with the assistance of an experienced metallurgical consultant. The most common causes of corrosive failure in stainless steels are localized attacks, in the form of intergranular corrosion, stress corrosion, pitting, and crevice corrosion. Localized corrosion attack also occurs quite often at areas contaminated because of improper handling (see below).

Design procedures for austenitic stainless steels will be the same as for carbon-steel vessels, modified by the technical properties of austenitic stainless steels.

Table 11.6.

STAINLESS STEEL ASTM TYPE	NOMINAL COMPOSITION	RELATIVE COST FACTOR
Ferritic		
405	12Cr–Al (0.08 C max. and 0.10–0.30 Al)	0.85
429	15Cr (0.12 C max.)	
430	17Cr (0.12 C max.)	0.90
446	27Cr (0.20 C max.)	
Martensitic		
410S	13Cr (0.03 C max.)	0.90
409	11Cr–Ti (0.12 C max.)	
410	13Cr (0.15 C max.)	0.80
Austenitic		
201	17Cr–4Ni–6Mn	0.85
302	18Cr–8Ni (0.15 C max.)	0.95
303	18Cr–8Ni (0.15 C max.)	1.04
304	18Cr–8Ni (0.08 C max.)	1.00
304L	18Cr–8Ni (0.03 C max.)	1.10
316	16Cr–12Ni–2Mo (0.08 C max.)	1.40
316L	16Cr–12Ni–2Mo (0.03 C max.)	1.55
317	18Cr–13Ni–3Mo (0.08 C max.)	2.01
317L	18Cr–13Ni–3Mo (0.03 C max.)	2.20
321	18Cr–10Ni–Ti (stabilized)	1.32
347	18Cr–10Ni–Cb (stabilized)	1.70
309	23Cr–12Ni (0.20 C max.)	1.56
309S	23Cr–12Ni (0.08 C max.)	1.70
310	25Cr–20Ni (0.25 C max.)	2.05
310S	25Cr–20Ni (0.08 C max.)	2.22

Since austenitic steels are used mainly for highly severe environmental service conditions, they are more often subject to stress-corrosion cracking than are carbon steels. Under such conditions they are sensitive to stress concentrations, particularly at higher temperatures (above 650°F) or at normal temperatures under cyclic conditions. Surface conditions have a great effect on fatigue strength and on corrosion resistance. Weld surfaces in contact with the operating fluid are often ground smooth. Welded connections should be designed with minimum stress concentrations. Abrupt changes, e.g., fillet welds, should be avoided, and butt welds should preferably be used and fully radiographed. The edge of the weld deposits should merge smoothly into the base metal without undercuts or abrupt transitions. Sound, uncontaminated welds are important. The welds should be located away from any structural discontinuities. Weld deposition sequences should be used that will minimize the residual stresses. Austenitic steels conduct heat more slowly than carbon steel, and expand and contract to a greater degree for the same temperature change. Thermal differentials, larger than for carbon steels, should be minimized by construction which permits free movement. These steels tend to warp and crack due to thermal stresses. All stainless steel parts should be designed to facilitate cleaning and self-draining without any crevices or spots, where dirt could accumulate and obstruct the access of oxygen to form the protective layer.

Fabrication and Handling. Since stainless steels exhibit the maximum resistance to corrosion only when thoroughly clean, preventive measures to protect cleaned surfaces should be taken and maintained during fabrication, storage, and shipping. Special efforts should be made at all times to keep stainless steel surfaces from coming into contact with other metals. For cleaning, only clean stainless steel wool and brushes should be used. If flame cutting is used, additional metal should be removed by mechanical means to provide clean, weldable edges. All grinding of stainless steels should be performed with aluminum oxide or silicon carbide grinding wheels bonded with resin or rubber, and not previously used on other metal. Proper identification and correct marking of the types of the material is important.

11.6. SELECTION OF STEELS FOR HYDROGEN SERVICE

Since many high-temperature petroleum refining processes are carried out under high partial pressures of hydrogen, the material selected for vessel construction for such service should safely withstand hydrogen attack, which can cause a deterioration of the material and subsequent failure.

The mechanism of hydrogen damage in steels is based on the fact, that the monatomic hydrogen diffuses readily through metal, whereas molecular hydrogen does not. Diffusion of atomic hydrogen through steels depends on a number of factors such as temperature, partial hydrogen pressure, time and composition of the material.

When atomic hydrogen is created by a chemical reaction *at intermediate temperatures* at the steel surface, for instance by hydrogen sulfide attack, it diffuses readily into the metal before it can form molecular hydrogen. If the hydrogen atoms enter the voids in the steel and then combine to form molecular hydrogen, they can no longer diffuse out.

The voids can be microscopic in size, gross laminations, or slag inclusions. If the diffusion of atomic hydrogen is sustained for a period of time, high pressure can build up, which can either crack the shell plate or, if the voids are under the surface, cause *blistering*. Dirty steels and hard steels with yield strength close to the ultimate strength are susceptible to this form of hydrogen attack. In the case of a steel with comparatively low yield strength there is sufficient ductility to withstand these internal local stress concentrations, although ferritic steels may suffer blistering if the hydrogen source is considerable.

At higher temperatures (over 600°F) the atomic hydrogen, in addition to blistering and cracking, can attack steel, particularly plain carbon steel, in the following manner. At elevated temperatures atomic hydrogen can be formed by dissociation of molecular hydrogen. The iron carbide (Fe_3C) in pearlite tends to decompose, with carbon atoms precipitating toward the grain boundaries. At the microscopic boundary voids, these carbon atoms combine with permeated atomic hydrogen to form methane gas (CH_4), which again cannot diffuse out of the steel. After the pressure has been built up high enough, intergranular cracking results, which is visible under the microscope. The material becomes spongy and embrittled, and the damage is permanent, since it cannot be reversed by any heat treatment. Notch toughness and fatigue strength are markedly reduced. Under the stresses imposed by operating conditions combined with any residual stresses, microcracks can propagate into a delayed fracture, with disasterous consequences.

Hydrogen embrittlement, a condition of low ductility in steel resulting from the absorption of atomic hydrogen in the metal lattice is only a temporary phenomenon. After the atomic hydrogen diffuses out of the metal, ductility is restored.

Preventing Hydrogen Damage by the Selection of Suitable Materials

Plain carbon and low-alloy steels are satisfactory for hydrogen service at low operating temperatures and high pressures or low pressures and high temperatures. The damage will be influenced by the cleanliness of the steel. Sound, inclusion-free welds are desirable in high-temperature hydrogen service. The reaction of hydrogen with the carbides in steels can be decreased or prevented by adding carbide-stabilizing elements such as chromium or molybdenum. Non–carbide forming elements such as nickel and silicon do not prevent internal hydrogen damage to steels. To provide adequate protection against hydrogen attack it is often necessary to use low-alloy steels containing chromium or molybdenum or both.

In selecting steels for petrochemical service the Nelson curves [87, 119], which show the operating limits for steels in hydrogen service, are strictly followed and extensively used. According to these curves carbon or low-alloy steels are not attacked by atomic hydrogen if the operating temperature and hydrogen partial pressure fall on or below or to the left of the curve applicable for the type of steel. A safety factor need not be applied to these curves. Maximum operating partial hydrogen pressure and temperature inside the vessel are used in selection of the material. In most material applications in the petrochemical industry, hydrogen sulfide corrosion will probably require a higher grade alloy steel than will hydrogen service alone.

All of the austenitic stainless steels resist hydrogen damage. However, atomic hydrogen still diffuses through them. Therefore, the backing material for stainless-steel-clad steel is selected as hydrogen attack resistant for hydrogen service, without taking any benefit for possible protection by the stainless steel layer. Hydrogen embrittlement can become important in martensitic steels; it is less important in ferritic and virtually unknown in austenitic stainless steels.

Design for Hydrogen Service

Special design combined with closely supervised fabrication techniques may be necessary for process equipment exposed to a hydrogen atmosphere at high pressures and temperatures. For hydrogen service as encountered in petroleum refineries a well designed vessel performs quite successfully. In practice, the designer should carefully avoid any stress concentrations, include in specifications additional tests for any laminar discontinuities in the plates and impurities in the welds, particularly in thicker plates (over 2 in. thick), and use normalized material with a minimum of residual stresses.

Figure 11.5 shows a fracture in a heat-exchanger channel shell due to the combined effect of built-up hydrogen pressure in the original plate lamination, differential thermal expansion between the ill fitted reinforcing pad and the shell (the air between them acting as a thermal barrier), and the probable residual stresses in the heat-affected zone between the welds. The cracking could have been prevented by eliminating plate lamination defects by means of an ultrasonic test (not required by the Code), making the distance between the flange and the reinforcing-pad welds larger than the Code required minimum, and by

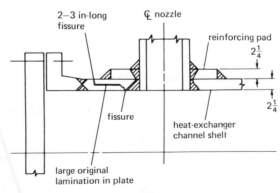

Material: A204 gr. B
Design temperature: 750°F
S.R., Full XR

Fig. 11.5.

blending the reinforcing pad fillet weld smoothly into the shell. Also, a nozzle with an integral reinforcement should have been considered.

11.7. ALUMINUM ALLOYS

In petrochemical plants aluminum alloys are not often used as construction materials for pressure vessels. In spite of many outstanding qualities (light weight, good corrosion resistance) they cannot match the steel alloys in price, strength above room temperature, reliability, and service life in heavily polluted atmospheres. Also, many welding shops are not equipped to handle the fabrication of all-aluminum-alloy pressure vessels.

If aluminum alloys are employed for low-temperature equipment such as storage tanks, alloy 3003, which is the least expensive and of good weldability and workability, is generally specified. If higher strength is required at temperatures below 150°F, the easily worked alloys 5083 or 5086 are used; at temperatures above 150°F, alloys 5454 or 6061 should be considered.

For cryogenic vessels aluminum alloys become more competitive with stainless steels and are more often used. There are no Code requirements for impact testing of wrought aluminum alloys down to -425°F, or for cast aluminum alloys down to -325°F. The alloys 5050-O, 5083-O (preferred), 5086-O, and 5454-O are most frequently used. On average the coefficient of thermal expansion α for aluminum alloys is about twice as large as that for steel ($\alpha = 12.9 \times 10^{-6}$ in./in./°F for alloy 3003), an important factor in the design of piping for cryogenic temperatures. The modulus of elasticity of aluminum alloys is only one third of that for carbon steel ($E = 10 \times 10^6$ at room temperature). Under the same loading conditions, an aluminum structure has three times the elastic deflection of a geometrically identical steel structure, and also about one-third the weight.

Aluminum surfaces which will be in contact with concrete or mortar after erection must be well protected by a protective coating, and contact of aluminum alloys with other metals must be prevented by insertion of nonmetallic separators for protection against electrolytic corrosion.

The strength of aluminum alloys can be increased by adding alloying elements, by heat treatment, or by cold working. Accordingly, aluminum alloys can be divided into two classes: heat-treatable and non–heat treatable alloys. The *non–heat treatable alloy group* includes high-purity aluminum and alloys in the 1000, 3000, 4000, and 5000 series. Their strength depends on the amount of cold working. The strength attained through cold work can be removed by heating in the annealing range.

The *heat-treatable alloy group* includes the alloys in the 2000, 6000, and 7000 series. Their strength can be increased by thermal treatment. High-temperature treatment is followed by quenching in water. The alloys become very workable but unstable: After an aging period of several days at room temperature, precipitation of constituents from the supersaturated solution begins and the alloys become considerably stronger. A further increase in strength can be attained by heating for some time. This is called *artificial aging* or *precipitation hardening*. Fusion welding always produces partial annealing in and near the weld zone in aluminum alloys, reduces the weld strength, and impairs corrosion resistance. The Code recognizes this by allowing the vessel designer to use the allowable stress values of –0 or –7 type alloys as welded material for welded constructions, which include practically all vessels. Fusion welding of the "strong" alloys is not recommended unless subsequent heat treatment is possible.

Aluminum Alloy Designation [97]

Wrought aluminum alloys are designated by a four-digit number, for instance 3003. The first digit indicates the alloy group, according to the major alloying elements, as follows:

1. 99+ percent pure aluminum,
2. copper,
3. manganese,
4. silicon,
5. magnesium,
6. magnesium and silicon,
7. zinc,
8. other elements.

The second digit indicates specific alloy modification of the original alloy. If the second digit is zero, it indicates the original alloy. The last two digits identify the specific aluminum alloy in the group. The letter, following and separated from the alloy designation number by a dash, designates the temper of the alloy:

-F: as fabricated, no special controls are applied (properties of F temper are not specified or guaranteed),
-O: annealed and recrystallized,
-H: cold worked, strain hardened (always followed by two or more digits),
-T: heat treated (followed by one or more digits).

Additional numbers following the letter, indicate degree of cold working or a specific heat treatment:

-H1: strain hardened only,

-H2: strain hardened, then partially annealed,
-H3: strain hardened, then stabilized by applying low-temperature thermal treatment,
-T4: solution heat treated and naturally aged to a stable condition,
-T6: solution heat treated and then artificially aged.

The second digit after the letter indicates the final degree of strain hardening: digit 2 is quarter-hard, 4 is half-hard, 6 is three-quarter hard, and 8 is full hard. For instance, -H18 alloy is strain hardened, full hard. Extra-hard tempers are designated by the digit 9. The third digit indicates a variation of the two-digit temper. The -H112 temper is generally considered a controlled F-temper with guaranteed mechanical properties.

Appendices

A1 Wind and Earthquake Maps

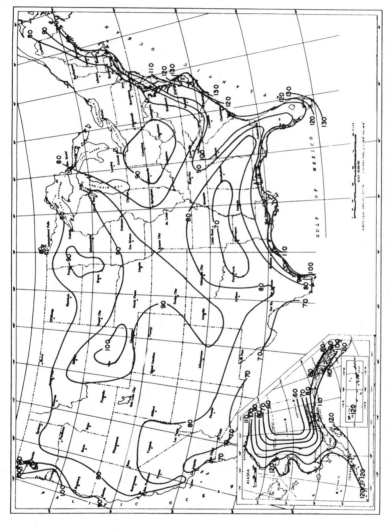

Fig. A1.1. Basic wind speed in miles per hour annual extreme fastest-mile speed 30 feet above ground, 100-year mean recurrence interval. (This material is reproduced with permission from American National Standard A58.1-1972 copyright 1972 by the American National Standards Institute, copies of which may be purchased from the American National Standards Institute at 1430 Broadway, New York, New York 10018.)

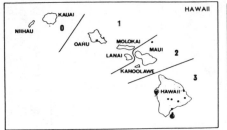
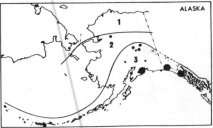

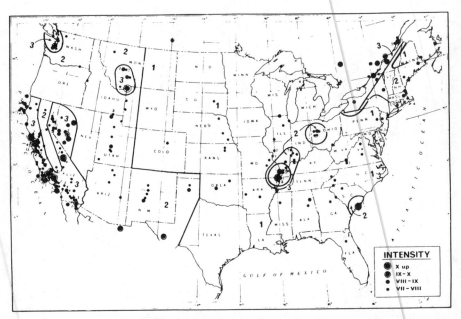

Fig. A1.2. Risk zones and damaging earthquakes of the United States through 1968. (This material is reproduced with permission from American National Standard A58.1-1972 copyright 1972 by the American National Standards Institute, copies of which may be purchased from the American National Standards Institute at 1430 Broadway, New York, New York 10018.)

A2 Geometric and Material Charts for Cylindrical Vessels

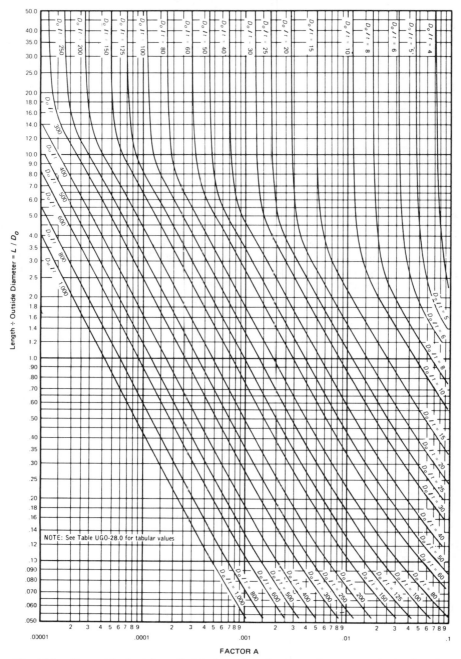

Fig. A2.1. Geometric chart for cylindrical vessels under external or compressive loading (for all materials). (Reproduced from the ASME Boiler and Pressure Vessel Code, Section VIII, Division 1, 1977 edition, by permission of the American Society of Mechanical Engineers.)

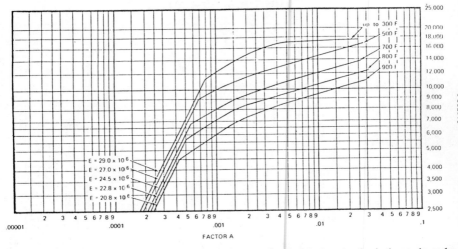

Fig. A2.2. Chart for determining shell thickness of cylindrical and spherical vessels under external pressure when constructed of carbon or low-alloy steels. (Reproduced from the ASME Boiler and Pressure Vessel Code, Section VIII, Division 1, 1977 edition, with permission of the American Society of Mechanical Engineers.)

A3 Skirt Base Details

Table A3.1.

B.C. o.d. shirt

f dia. bolt hole

k c t_r t_b

12

g dia. bolt hole

a t_{sk} 2 in. min.

dimension a = socket wrench radius plus $\frac{3}{8}$ in.

BOLT SIZE	a	MIN. n	c	k	f DIA.	g DIA.
1	$1\frac{3}{4}$	3	$2\frac{1}{4}$	$\frac{1}{2}$	$1\frac{1}{4}$	$1\frac{1}{2}$
$1\frac{1}{8}$	$1\frac{7}{8}$	$3\frac{1}{4}$	$2\frac{3}{8}$	$\frac{1}{2}$	$1\frac{3}{8}$	$1\frac{5}{8}$
$1\frac{1}{4}$	2	$3\frac{1}{2}$	$2\frac{1}{2}$	$\frac{1}{2}$	$1\frac{1}{2}$	$1\frac{3}{4}$
$1\frac{3}{8}$	$2\frac{1}{8}$	4	$2\frac{3}{4}$	$\frac{5}{8}$	$1\frac{5}{8}$	$1\frac{7}{8}$
$1\frac{1}{2}$	$2\frac{1}{4}$	$4\frac{1}{2}$	3	$\frac{5}{8}$	$1\frac{3}{4}$	2
$1\frac{5}{8}$	$2\frac{3}{8}$	$4\frac{3}{4}$	$3\frac{1}{4}$	$\frac{5}{8}$	$1\frac{7}{8}$	$2\frac{1}{8}$
$1\frac{3}{4}$	$2\frac{5}{8}$	5	$3\frac{3}{8}$	$\frac{3}{4}$	2	$2\frac{1}{4}$
$1\frac{7}{8}$	$2\frac{5}{8}$	5	$3\frac{3}{4}$	$\frac{3}{4}$	$2\frac{1}{8}$	$2\frac{3}{8}$
2	$2\frac{3}{4}$	$5\frac{1}{4}$	$3\frac{5}{8}$	$\frac{3}{4}$	$2\frac{1}{4}$	$2\frac{1}{2}$
$2\frac{1}{8}$	$2\frac{7}{8}$	$5\frac{3}{4}$	$3\frac{3}{4}$	$\frac{3}{4}$	$2\frac{3}{8}$	$2\frac{5}{8}$
$2\frac{1}{4}$	3	$5\frac{3}{4}$	$3\frac{7}{8}$	1	$2\frac{1}{2}$	$2\frac{3}{4}$
$2\frac{1}{2}$	$3\frac{3}{8}$	$6\frac{3}{8}$	$4\frac{1}{8}$	1	$2\frac{3}{4}$	3
$2\frac{3}{4}$	$3\frac{5}{8}$	$6\frac{3}{4}$	$4\frac{3}{8}$	$1\frac{1}{4}$	3	$3\frac{1}{4}$
3	$3\frac{7}{8}$	$7\frac{1}{4}$	$4\frac{5}{8}$	$1\frac{1}{4}$	$3\frac{1}{4}$	$3\frac{1}{2}$

Table A3.2.

anchor bolt opening in skirt

c e L t_b

t_{sk}

m—pipe size sch. 160

f dia. bolt hole

Mean dia. skirt = mean dia. base ring = bolt circle

BOLT SIZE	L	MIN. b	e	c	f DIA.	m
1	3	3	3	$3\frac{1}{4}$	$1\frac{1}{4}$	2
$1\frac{1}{8}$	3	3	3	$3\frac{1}{2}$	$1\frac{3}{8}$	2
$1\frac{1}{4}$	3	3	$3\frac{3}{8}$	$3\frac{3}{4}$	$1\frac{1}{2}$	2
$1\frac{3}{8}$	3	4	$3\frac{3}{4}$	4	$1\frac{5}{8}$	$2\frac{1}{2}$
$1\frac{1}{2}$	4	4	$4\frac{1}{8}$	$4\frac{1}{4}$	$1\frac{3}{4}$	$2\frac{1}{2}$
$1\frac{5}{8}$	4	4	$4\frac{3}{8}$	$4\frac{1}{2}$	$1\frac{7}{8}$	$2\frac{1}{2}$
$1\frac{3}{4}$	5	5	$4\frac{3}{4}$	$4\frac{3}{4}$	2	3
$1\frac{7}{8}$	5	5	$5\frac{1}{8}$	5	$2\frac{1}{8}$	3
2	6	5	$5\frac{1}{2}$	$5\frac{1}{4}$	$2\frac{3}{8}$	3
$2\frac{1}{4}$	7	5	$5\frac{7}{8}$	$5\frac{3}{4}$	$2\frac{5}{8}$	3
$2\frac{1}{2}$	8	6	6	$6\frac{3}{4}$	$2\frac{7}{8}$	4
$2\frac{3}{4}$	9	6	$6\frac{3}{8}$	$6\frac{3}{4}$	$3\frac{1}{8}$	4
3	10	6	7	7	$3\frac{1}{2}$	5

A4 Sliding Supports for Vertical and Horizontal Vessels

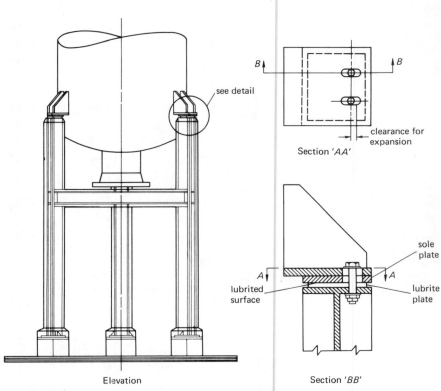

see detail

B ↑   ↑ B

clearance for expansion

Section 'AA'

sole plate

A ↓   ↓ A

lubrited surface

lubrite plate

Elevation

Section 'BB'

Fig. A4.1. Self-lubricating sliding base plates for vertical vessels. (By permission of Litton-Merriman Inc.)

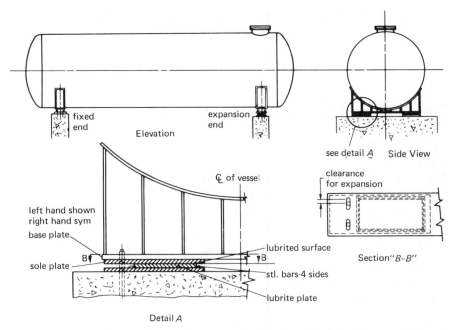

Fig. A4.2. Self-lubricating sliding base plates for horizontal vessels. (By permission of Litton-Merriman Inc.)

A5 Glossary of Terms Relating to the Selection of Materials

Aging a change in properties in a metal or alloy that generally occurs slowly at room temperature (natural aging) and more rapidly at higher temperatures (artificial aging).

Alloy an intentional combination of two or more substances, of which at least one is a metal, which exhibits metallic properties. It can be either a mixture of two types of crystalline structures or a solid solution.

Amorphus noncrystalline.

Baushinger effect the application of a tensile force to a polycrystalline metal in the elastic range raises the proportional limit of the metal but lowers its compressive proportional limit.

This phenomenon is attributed to the fact that under tensile stress some yielding occurs in the most unfavorably oriented crystals or grains. When the tensile force is removed, some residual stresses result.

If a tensile force is applied for a second time, the residual stresses must be overcome and the proportional limit is raised. However, the proportional limit in compression is lowered.

Similarly, the stressing of a metal into the inelastic range results in the increase of the yield strength for subsequent loadings in the same direction.

Carburization occurs frequently on the inner surface of furnace tubes exposed to hydrocarbon streams. The carbide formed embrittles the steel and is the cause of tube failures. Generally, austenitic stainless steels can tolerate a greater increase in carbon than can ferritic stainless steels without affecting ductility.

Caustic dip a strongly alkaline solution into which metals are immersed for neutralizing acid or removing organic materials such as grease.

Corrosion deterioration of a material (usually metal) due to its reaction with the environment.

Corrosion processes may be classified basically into two types:

1. *Chemical corrosion* (dry) is a direct chemical attack by a particular chemical in the environment on the metal. The intensity of the attack usually increases with rising temperature.

The various reactions between metals and chemicals and the corrosion rates are generally well known. Typical examples are direct oxidation of iron.

$$(3Fe + 2O_2 = Fe_3O_4) \text{ or sulfidic corrosion } (SO_2 + Fe = FeS + O_2).$$

2. *Electrochemical corrosion* (galvanic, wet) is due to a potential difference which develops between two points (anode and cathode) in electrical contact and in the presence of an electrolyte (any liquid, for instance water, which contains ions).

A potential difference between two points or areas on the metal surface may be due to the variation in composition of one metal or of electrolyte (areas of different concentration of the absorbed oxygen) or due to the presence of two different metals.

The area where the metal atoms dissolve more readily into solution to become ions, and which therefore has less corrosion resistance, becomes the anode, and the more corrosion-resistant metal area becomes the cathode, where the electric current enters the metal from the solution and no corrosion occurs.

The passage of a current through an electrolytic conductor is always accompanied by reactions at the electrodes. The typical example is the rusting of iron. The entire reaction can be described as follows (see Fig. A5.1):

At the anode, Fe metal on the surface looses two electrons on contact with water and dissolves as ferrous ion (Fe^{2+}) in the water. It can be further oxidized by dissolved atmospheric oxygen to ferric ion Fe^{3+}:

$$Fe \longrightarrow Fe^{2+} + 2e^-$$
$$Fe^{2+} \longrightarrow Fe^{3+} + e^-.$$

The released electrons flow through the metal to the cathode.

At the cathode, hydroxyl ions $(OH)^-$ are formed according to the equation

$$2H_2O + O_2 \text{ (in solution from air)} + 4e^- = 4(OH)^-.$$

In the water, the negative hydroxyl ion combines with the positive ferric ion to form ferric hydroxide, which is insoluble in water and precipitates as rust:

$$Fe^{3+} + 3(OH)^- \longrightarrow Fe(OH)_3.$$

There are other possible reactions in addition to those described.

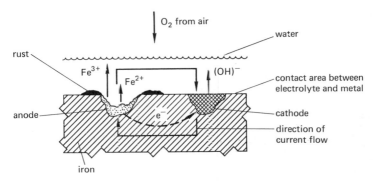

Fig. A5.1. Rusting of iron.

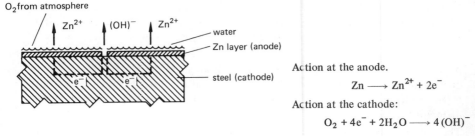

Action at the anode.

$$Zn \longrightarrow Zn^{2+} + 2e^-$$

Action at the cathode:

$$O_2 + 4e^- + 2H_2O \longrightarrow 4(OH)^-$$

Fig. A5.2.

A common protection of iron parts against galvanic corrosion is a thin layer of zinc formed by dipping the iron parts into hot liquid zinc; for instance, galvanized (zinc-plated) pipes. Since Zn is anodic with respect to steel, it will provide protection even if the zinc layer is not continuous, as shown in Fig. A5.2.

On the other hand, if tin is used for the protective layer, it must be continuous, since tin, being more stable than iron, behaves as a cathode with respect to steel. If the tin layer is disrupted (scratched), steel will corrode as shown in Fig. A5.3. According their mode of resistance, corrosion-resistant metals and alloys usually can be classified into two groups: those able to maintain their metallic structure in direct contact with the environment (noble metals) and those with the ability to form a firm continuous (passive) films and so become isolated from the electrolyte. A typical example of corrosion resistance by a protective surface film is chromium steel. Cr is anodic with respect to steel however, it combines with oxygen in solution and forms a passive film in the metal surface that prevents any further corrosion of the metal. Various types of corrosion associated with vessel failures are listed below, with short descriptions.

Stress corrosion cracking is the failure by cracking (brittle failure) of a material due to the combined effects of galvanic corrosion and tensile stress. It occurs most often at points of high stress concentration or high residual stresses.

Localized corrosion is needed to initiate stress corrosion cracking. The crack is propagated by simultaneous corrosion of the metal and tearing of any protective film that may form at the notch tip. The crack will gradually increase until the remaining cross section of the part can no longer resist the load and failure occurs.

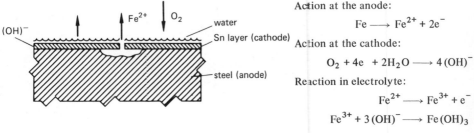

Action at the anode:

$$Fe \longrightarrow Fe^{2+} + 2e^-$$

Action at the cathode:

$$O_2 + 4e^- + 2H_2O \longrightarrow 4(OH)^-$$

Reaction in electrolyte:

$$Fe^{2+} \longrightarrow Fe^{3+} + e^-$$

$$Fe^{3+} + 3(OH)^- \longrightarrow Fe(OH)_3$$

Fig. A5.3.

Stress corrosion may be minimized by selecting proper material to prevent pitting, by stress relieving to reduce residual stresses, and by proper design without any stress raisers. Alloys with lower carbon content are tougher and more stress-corrosion resistant than those with higher carbon content.

Corrosion fatigue is failure by stress corrosion under cyclic stresses.

Corrosion pitting is localized galvanic corrosion, also described as concentration cell corrosion. It may result from differences in oxygen concentration of the electrolyte in one spot (pit) on the metal surface acting as an anode and outside the pit, acting as a cathode. The spots with less oxygen act as anodes. Susceptibility to pitting corrosion is increased by surface defects like scratches.

This type of corrosion is quite frequently found in ferrous or nonferrous metals. In stainless steels, pitting failures are caused by chloride ions, which tend to break through the passive film. Some materials are more resistant to pitting than others. For instance, the addition of 2 percent molybdenum to 18-8 stainless steel type 316 results in a very large increase in resistance to corrosion pitting.

Caustic embrittlement is the cracking of metal under tensile stress and in contact with an alkaline solution.

General corrosion is corrosion, either chemical or galvanic, which proceeds uniformly over the entire exposed surface.

Intergranular corrosion is a localized galvanic corrosion at the grain boundaries of sensitized metal, particularly stainless steels. The grain boundaries are usually anodic to the grains.

Crystal a solid composed of atoms (or molecules) arranged in a certain geometric pattern (crystal system) which is periodic in three dimensions.

The elastic properties (E and G) of a crystal vary to a considerable extent depending on the orientation of the crystal axis with respect to the induced stress.

Metals usually have a microcrystalline structure. They consist of large number of minute crystals, called grains, whose principal axes have random orientations, so they can be considered as isotropic materials, that is, materials with the same physical properties in all directions.

Crystal system one of seven groups into which all crystals may be divided: cubic, tetragonal, orthorhombic, monoclinic, triclinic, hexagonal, and rhombohedral. The names of the groups describe the arrangement of the atoms.

The cubic system is subdivided by unit cells (space lattices): simple cubic, body-centered cubic (BCC) as in the α-iron crystals, and face-centered (FCC) as in γ-iron crystals.

Metals with FCC structure do not generally exhibit loss of impact strength at very low temperatures.

Decarburization loss of carbon from a ferrous alloy, usually at the surface, as a result of heating in an environment that reacts with the carbon at the surface.

Deoxidized copper copper from which cuprous oxide has been removed by adding a deoxidizer (such as phosphorus) to the molten metal.

Ductility ability of a metal to deform in the plastic range without fracturing under stress.

Endurance limit stress below which an alternating stress will not cause a failure. The number of cycles N is infinite—usually not less than 10^6 in tests.

Erosion destruction of a metal by the abrasive action of a liquid or vapor or of solid particles suspended in the operating liquid or vapor.

Fatigue life finite number of stress cycles N which can be sustained under operating conditions.

Fatigue strength stress at which failure occurs at some definite number of cycles N. If measured in a corrosive environment it is called *corrosive fatigue strength*. Endurance limit S_n is then the highest cyclic stress the metal can be subjected to indefinitely (for steel, about 10^6 cycles) without fracture. (The stresses a metal can tolerate under cyclic loadings are much lower than under static loadings.)

Galling seizure or welding of parts in close contact under pressure during relative motion.

Grain an individual crystal in a polycrystalline metal or alloy with distinct boundaries. The *grain boundary* is the interface separating two grains, where the orientation of the lattice changes from that of one grain to that of another. Between two boundaries there is a *transition zone*, where the atoms are not aligned exactly with either grain. The transition zone is considered to be of high energy and stronger than grain itself.

Grain boundary sensitization occurs in stainless steels in solution-annealed condition during any heating in the 800–1600°F temperature range. See Chapter 11 for details.

Grain size is usually specified by ASTM index numbers, based on the following formula: $n = 2^{N-1}$, where N is the ASTM index number and n is the number of grains per one square inch at 100 times magnification. n ranges from 1 to 128; N ranges from 1 to 8. Steels with $N = 1$–5 are considered *coarse grained*; steels with $N = 6$–8 are *fine grained*. For example if $N = 5$ and $n = 16$, and assuming the grain to be roughly square in cross section, its average actual side dimension would be $(1 \times 1/16 \times 100 \times 100)^{1/2} = 0.0025$ in. Grain size may have a marked effect on creep and creep rupture life of the material selected.

Generally, *fine-grained steels* have a lower transition temperature than do those of coarse grain. Steels with finer austenitic grain also have a greater toughness (i.e., ability to absorb energy during plastic deformation), lower hardenability, and lower internal stresses.

Coarse-grained steels excel in rupture strength with increasing test time and temperatures. A coarse-grained high-temperature alloy has greater creep strength than a fine-grained alloy. However, larger grain size results in embrittlement and lowered ductility, since larger grains allow more penetration of carbon, nitrogen, or oxygen (for instance in welding), which form brittle phases at grain boundaries. Low ductility is usually considered an indication of poor fatigue properties.

Grain growth increase in size of grain by reduction in number of grains. Grain growth will take place at higher temperatures during normal use. Loss of ductility due to grain growth limits the metal usefulness.

Graphitization precipitation of carbon in the form of graphite at grain boundaries, as occurs if carbon steel is in service long enough above 775°F, and C–Mo steel above 875°F. Graphitization appears to lower steel strength by removing the strengthening effect of finely dispersed iron carbides (cementite) from grains. Fine-grained, aluminum-killed steels seem to be particularly susceptible to graphitization.

Graphitization during solidification is called *primary*; during heating, *secondary*. In welds, there is a possibility of graphite formation in the affected zone.

Graphitizer material that increases the tendency of iron carbide (Fe_3C) to break down into iron and graphite.

Inclusion nonmetallic material (oxides, silicate, etc.) held mechanically in a metallic matrix as unintentional impurities.

Heat-affected weld zone a portion of the base metal adjacent to the weld which has not been melted, but whose metallurgical microstructure and mechanical properties have been changed by the heat of welding, usually for the worse.

Heat treatment heat treating operation performed either to produce changes in mechanical properties of the material or to restore its maximum corrosion resistance. There are three principal types of heat treatment; annealing, normalizing, and post-weld heat treatment.

Annealing is heating to and holding at a suitable temperature and then cooling at a suitable rate, for such purposes as reducing hardness, producing desired microstructure, etc.

Normalizing is heating steel to above its critical temperature and cooling in still air.

Post-weld heat treatment (PWHT) is uniform heating of a weldment to a temperature below the critical range to relieve the major part of the residual stresses, followed by uniform cooling in still air. Heating reduces the metal yield strength, and any locked-in residual stress greater than the depressed yield strength will start to dissipate by plastic displacement. The maximum remaining residual stress will be equal to the depressed yield strength of the metal at the PWHT temperature.

Preheat treatment usually heating the base metal before welding to decrease the temperature gradient between the weld metal, heat-affected zone, and the base metal. Preheat treatment has been found very effective in avoiding weld or heat-affected-zone cracking.

Quenching rapid cooling of a metal from above the critical temperature, with change in macrostructure.

Solution heat treatment consists of heating an alloy to a suitable temperature, holding at that temperature long enough to allow one or more constituents to enter in solid solution within the grains of the base metal, and then cooling rapidly to hold this solution at lower temperatures.

Heat-treatable alloy an alloy which may be strengthened by a suitable heat treatment.

Hydrogen embrittlement low ductility in metals resulting from the absorbtion of the atomic hydrogen.

Macrostructure structure of metal as seen on a polished, etched surface by the eye or under low magnification (not over 10 diameters).

Microstructure structure of metal as seen on a polished etched surface under high magnification (over 10 diameters).

Nil-Ductility temperature (NDT) a temperature at which, under a light load, crack initiation and propagation results in a brittle fracture.

Nodular cast iron cast iron that has been treated while molten with a master alloy containing an element such as Mg to give primary graphite in spherulitic form.

Notch brittleness susceptibility of a material to brittle fracture at points of stress concentration. In a notch tensile test the material is said to be *notch brittle* if the notch strength is less than the tensile strength. Otherwise, it is said to be *notch ductile*.

Notch sensitivity measure of the reduction in the strength of a metal caused by the presence of a stress concentration. Values can be obtained from static, impact, or fatigue tests.

Notch toughness ability of a material to resist brittle fracture under conditions of high stress concentration, such as impact loading in the presence of a notch. This is an important property for construction materials intended for low-temperature service.

Oxidation or scaling of metals occurs at high temperatures and access of air. Scaling of carbon steels from air or steam is negligible up to 1000°F. Chromium increases scaling resistance of carbon steels. Decreasing oxidation resistance makes austenitic stainless steels unsuitable for operating temperatures above 1500°F.

Passivation changing of the chemically active surface of a metal to a much less reactive state, usually by forming an impervious layer of oxide or other compound.

Permeability a property of a material which permits passage of gas through the molecular structure.

Phase a physically homogeneous, mechanically separable, distinct portion of a material system.

Polycrystalline characteristic of an aggregate composed of usually large number crystals or grains (polygranular). Crystals have dissimilar orientation so that grain boundaries are present.

Precipitate to separate out a soluble substance from a liquid or solid solution.

Precipitation hardening increase in strength and loss of ductility and toughness during elevated-temperature service, due to compound formed at elevated temperatures.

Proof test the load inducing the stress at which a material shows no or a specified permanent elongation when the stress is removed.

Recrystallization change of a plastically deformed crystalline structure into a more perfect, unrestrained crystalline structure exhibiting higher softness. Recrystallization a high temperature is called *annealing* (see *Heat treatment*), and the temperature of marked softening is called the *recrystallization temperature.*

Refractory a material of very high melting point with properties that make it suitable for such uses as high-temperature lining.

Stabilizer any material that increases the tendency of carbon to remain as iron carbide or in solid solution within steel crystals and retards graphitization.

Steel crystalline phases

Alpha iron (α-Fe) is the solid phase of pure iron stable below $1670°$F, with body-centered cubic latice (BCC). It is ferromagnetic at temperatures under $1418°$F.

Austenite is a solid solution of carbon or other elements in γ-iron. The solute is usually assumed to be carbon, unless otherwise noted. It is soft and ductile, with considerable tensile strength. On slow cooling, austenite tends to decompose into ferrite and cementite. At room temperature it can be present only in alloy steels.

Cementite, loosely referred to as iron carbide (Fe_3C), is characterized by an orthorombic crystal structure. It is the hardest constituent in carbon steel, and is nonductile. Its presence in steel increases strength. Some atoms in the cementite latice may be replaced by manganese or chromium in alloys steels. Cemetite is in metastable equilibrium and tends to decompose into iron and graphite at higher temperatures.

Delta iron (δ-Fe) is the stable form of pure iron from $2540°$F to $2802°$F (melting point). It is of body-centered cubic crystalline structure. It has no practical application at the present time.

Ferrite is a solid solution of one or more elements in α-iron. The solute is generally assumed to be carbon unless otherwise noted. It exist in two forms: α-ferrite; and at temperature above $2540°$F, δ-ferrite.

Gamma iron (γ-Fe) is a solid, nonmagnetic phase of pure iron which is stable at temperatures between 1670°F and 2540°F and possesses the face-centered cubic latice (FCC).

Ledeburite is a eutectic mixture of austenite and cementite in metastable equilibrium formed on rapid cooling in iron–carbide alloys with carbon content more than two percent and less than 6.67 percent.

Martensite is a metastable phase in steel, hard and brittle, with needlelike appearance, formed by transformation of austenite below a certain temperature. Martensite is produced during quenching.

Pearlite is a lamellar aggregate of ferrite and cementite in alternate platelets formed by transformation of austenite of eutectoid composition on slow cooling. It is a microstructure formed in iron and carbon alloys.

Sigma phase forms when ferritic and austenitic stainless steels with more than 16.5 percent of chromium are heated at a prolonged period of time between 1100°F and 1700°F. It is a complex, nonmagnetic, intermetallic Cr–Fe compound, very hard and brittle at room temperature. A large percentage of sigma phase reduces the ductility, toughness, and also corrosion resistance of steel. It can be transformed into austenite and ferrite by suitable heating and quenching.

Sigma-phase precipitation occasionally occurs in tube supports or hangers made of 309 or 310 stainless steels in the radiant section of petroleum refinery furnaces. During shutdown for replacement of defective tubes, there supports become exceedingly brittle and frequently break when handled carelessly by maintenance crews.

Strain hardening if steel is stretched beyond the yield point into the plastic region, some increase in stress is required to produce an additional increase in strain. The phenomenon of strain hardening, by plastic deformation (as in cold working below the recrystallization temperature) improves the mechanical properties of steel, except that ductility is reduced. The hardening disappears if the material is subjected to annealing temperatures.

Stress range the algebraic difference between the maximum and minimum stress in one cycle.

Sulfide corrosion cracking stress corrosion cracking in sulfur-containing environments. The corrosive agent is usually hydrogen sulfide.

Thermal fatigue the development of cyclic thermal gradients producing high cyclic thermal stresses and subsequent local cracking of material.

Thermal shock the development of a steep thermal gradient and accompanying high stresses within a structure, as might occur when the surface of a cool metal part is heated rapidly.

Toughness ability of a material to absorb energy during plastic deformation.

Transition temperature (T.T.) the temperature at which the material changes from ductile to brittle fracturing. Since brittle fracture requires much less energy than shear–type fracture, a marked drop in absorbed energy occurs at a fracture below the transition temperature. For any service, the steel with lower T.T. is usually preferred, since it offers additional safety.

Ultimate strength the maximum load per unit of original cross-sectional area which a test specimen of a material can sustain before fracture, usually in single tension or compression.

Yield strength the load per unit of original cross-sectional area on a test specimen at which a marked increase in deformation occurs without the increase of load. A part of the deformation becomes permanent.

Weld structure the microstructure of a weld deposite and the heat-affected zone.

A6 Standard Specifications Pertaining to Materials

ASTM SPECIFICATIONS

A20 General Requirements for Delivery of Steel Plates for Pressure Vessels. Supplementary Requirements S1 to S13 when specified.
A262 Detecting Susceptibility to Intergranular Attack in Stainless Steels.
A370 Mechanical Testing of Steel Products.
A380 Descaling and Cleaning Stainless Steel Surfaces.
A435 Longitudinal Wave Ultrasonic Inspection of Steel Plates for Pressure Vessels.
A480 General Requirements for Delivery of Flat Rolled Stainless and Heat Resisting Steel Plate, Sheet and Strip.
A577 Ultrasonic Shear Wave Inspection for Steel Plates.
A593 Charpy V Notch Testing Requirements for Steel Plates for Pressure Vessels.
E7 Standard Definitions of Terms Relating to Metallography.
E21 Short-Time Elevated Temperature Tension Test of Material.
E94 Radiographic Testing.
E109 Dry Powder Magnetic Particle Inspection.
E112 Estimating the Average Grain Size of Metals.
E114 Ultrasonic Testing.
E139 Conducting Creep and Time-for-Rupture Tests for Materials.
E142 Controlling Quality of Radiographic Testing.
E165 Liquid Penetrant Inspection.
E186 Reference Radiographs for Heavy Walled (2 to $4\frac{1}{2}$ in. thick) Steel Castings.
E208 Conducting Drop Weight Test to Determine Nil Ductility Transition Temperature of Ferritic Steels.
E280 Reference Radiographs for Heavy Walled ($4\frac{1}{2}$ to 12 in.) Steel Castings.

AWS SPECIFICATIONS

A5.1 Specification for Mild Steel Covered Arc Welding Electrodes.
A5.4 Specification for Corrosion Resisting Chromium and Chromium Nickel Steel Covered Welding Electrodes.

ANSI SPECIFICATIONS

B1.1 Unified Screw Thread for Screws, Bolt, Nuts and Other Threaded Parts.
B18.2.1 Square and Hex. Bolt and Screws, Including Hex. Cap Screws and Lag Screws (heavy series).
B18.2.2 Square and Hex. Nuts (heavy series).
B16.5 Steel Pipe Flanges and Flanged Fittings.
B46.1 Surface Texture.

A7 Flanges

Most of the inspection openings or nozzles on vessels are provided with circular-type standard flanges for quick, easy disassembly of closing covers or connected piping. Only special flanges are designed or existing flanges are checked for maximum working pressures.

According to the available methods for solving the stresses in the flange connections under the pressure, flanges can be classified under four main categories, as described below.

1. *Flanges with a ring-type gasket inside the bolt circle* and with no contact outside the bolt circle under internal or external pressure. Flanges in this category are constructed of forged steel or forged low-alloy steel with a raised face and spiral-wound, metal–asbestos filled gasket. They are the most commonly used flanges. According how they are attached to the shell, they can be further subclassified as (a) *integral flanges*, such as welding neck flanges, which are welded or forged with the connection shell, forming a continuous structure, or (b) *loose flanges*, such as slip-on flanges or ring flanges, which are either not attached to the shell at all or attached by welds which are not strong enough to carry the load from the flange to shell and to form an integral part with the connection shell. The basic flange design was established in ref. 116 and is incorporated in the Code. The derivation of the design can also be found in refs. 44 and 117. Reference 110 is generally used in selecting the dimensions for a desired flange. Both types can be designed with different types of facings adjusted to the gasket. Slip-on flanges are used for lower pressures ($\leqslant 150$ psi) and where no temperature expansion stresses from connected piping are anticipated.

2. *Flanges with flat face and metal-to-metal contact inside and outside the bolt circle, with self-sealing gasket.* The basic design is derived in refs. 113 and 114, and the method is incorporated in the Code.

3. *Flanges with flat face and full-face gasket.* This type of the flange is used only for lower operating pressures. Design is as given in ref. 111.

4. *Special flanges.* These include large-diameter flanges (over 6 ft), flanges for high design pressures (over 1,500 psi), and flanges with special gaskets (double cone, lens, etc.), as described in ref. 112.

A8 References

1. ASME Boiler and Pressure Vessel Code, Section VIII. Pressure Vessels–Division 1. ASME, New York.
2. ASME Boiler and Pressure Vessel Code, Section VIII. Pressure Vessels–Division 2, Alternative Rules. ASME, New York.
3. ANSI A58.1-1972. Building Code Requirements for Minimum Design Loads in Buildings and Other Structures. ANSI, New York, 1972.
4. ANSI A58.1-1955. Requirements for Minimum Design Loads in Buildings and Other Structures. ANSI, New York, 1955.
5. Uniform Building Code. International Conference of Building Officials, 5360 South Workman Mill Road, Whittier, Cal. 90601, 1973.
6. Czerniak, E. Foundation design guide for stacks and towers. *Hydrocarbon Processing*, June 1969.
7. *Design of Piping Systems*. M. W. Kellogg Co., 1956.
8. Markl, A. R. C. Piping flexibility analysis. In *Pressure Vessel and Piping Design, Collected Papers,* ASME, New York, 1960.
9. ANSI B31.3 *Petroleum Refinery Piping*. ASME, New York, 1973.
10. *Design Criteria of Boilers and Pressure Vessels*. ASME, New York, 1969.
11. Osgood, Carl C. *Fatigue Design*. John Wiley & Sons, New York, 1970.
12. *Steels for Elevated Temperatures Service*. U.S. Steel Co., 1969.
13. Langer, B. F. Design values for thermal stresses in ductile materials. In *Pressure Vessel and Piping Design, Collected Papers,* ASME, New York, 1960.
14. Den Hartog, J. P. *Advanced Strength of Materials*. McGraw-Hill Book Co., New York, 1962.
15. Timoshenko, S. P. *Strength of Materials, Parts I and II*. 3d ed. D. Van Nostrand Co., Princeton, N.J., 1956.
16. Harvey, J. F. *Pressure Component Construction*. Van Nostrand Reinhold Co., New York, 1980.
17. Baker, E. H., L. Kovalevsky, and F. L. Rish. *Structural Analysis of Shells*. McGraw-Hill Book Co., New York, 1972.
18. Timoshenko, S. P., and S. Woinowsky-Krieger. *Theory of Plates and Shells*. McGraw-Hill Book Co., New York, 1959.
19. Roark, R. J., and W. C. Young. *Formulas for Stress and Strains*. 5th ed. McGraw-Hill Book Co., New York, 1975.
20. Kraus, H. *Thin Elastic Shells*. John Wiley & Sons, New York, 1967.
21. Ugural, A. C., and S. K. Fenster. *Advanced Strength and Applied Elasticity*. American Elsevier Publishing Co., New York, 1975.
22. Zick, L. P., and A. R. St. Germain. Circumferential stresses in pressure vessel shells of revolution. ASME Paper No. 62-PET-4.
23. Flügge, W. *Stresses in Shells*. Springer-Verlag, New York, 1960.
24. Esztergar, E. P., and H. Kraus. Analysis and design of ellipsoidal pressure vessel heads. ASME Paper No. 70-PVP-26.
25. Saunders, H. E., and D. F. Windenburg. Strength of thin cylindrical shells under external pressure. In *Pressure Vessel and Piping Design, Collected Papers*. ASME, New York, 1960.

26. Bergman, E. O. The new type Code chart for the design of vessels under external pressure. In *Pressure Vessel and Piping Design, Collected Papers*. ASME, New York, 1960.

27. Bergman, E. O. The design of vertical pressure vessels subjected to applied forces. In *Pressure Vessel and Piping Design, Collected Papers*, ASME, New York, 1960.

28. Lackman, L., and J. Penzien. Buckling of circular cones under axial compression. ASME Paper No. 60-APM-17.

29. Shield, R. T., and D. C. Drucker. Design of thin walled torispherical and toriconical pressure vessel heads. *J. Appl. Mech.*, 28, June 1961.

30. Thurston, G. A., and A. A. Holston. Buckling of cylindrical end closures by internal pressure. NASA CR-540, 1966.

31. Gallety, G. D. Elastic and elastic–plastic buckling of internally pressurized 2:1 ellipsoidal shells. *ASME J. Press. Vessel Technol.*, November 1978.

32. Mason, A. H. Find stiffener ring moment by nomograph. *Hydrocarbon Processing and Petroleum Refiner*, February 1962.

33. Siegel, M. I., V. L. Maleev, and J. B. Hartman. *Mechanical Design of Machines*. International Textbook Co., Scranton, Penna., 1965.

34. *Manual of Steel Construction*. AISC, New York, 1970.

35. Singer, F. L. *Strength of Materials*. 2nd ed. Harper & Row, New York, 1962.

36. Den Hartog, J. P. *Mechanical Vibrations*. McGraw-Hill Book Co., New York, 1956.

37. Rouse, H. *Fluid Mechanics for Hydraulic Engineers*. Dover Publications, New York, 1961.

38. Roehrich, L. R. Torquing stresses in lubricated bolts. *Machine Design*, June 1967.

39. Dann, R. T. How much preload for fasteners? *Machine Design*, August 1975.

40. Marshall, V. O. Foundation design handbook for stacks and towers. *Petroleum Refiner*, May 1958.

41. Nelson, J. G. Use calculation form for tower design. *Hydrocarbon Processing*, June 1963.

42. Brummerstedt, E. F. Design of anchor bolts for stacks and towers. *National Petroleum News*, January 5, 1944.

43. Gartner, A. J. Nomograms for the solution of anchor bolt problems. *Petroleum Refiner*, July 1951.

44. Brownell, L. E., and R. H. Young. *Process Equipment Design: Vessel Design*. John Wiley & Sons, New York, 1956.

45. Singh, K. P. Design of skirt mounted supports. *Hydrocarbon Processing*, April 1976.

46. Roark, R. J. *Formulas for Stress and Strain*. 4th ed. McGraw-Hill Book Co., New York, 1965.

47. Dickey, W. L., and G. B. Woodruff. The vibration of steel stacks. *ASCE Proceedings*, 80, November 1954.

48. DeGhetto, K., and W. Lang. Dynamic stability design of stacks and towers. *ASME Transactions*, 88, November 1966.

49. Zorrila, E. P. Determination of aerodynamic behavior of cantilever stacks and towers of circular cross section. ASME Paper No. 71-PET-35.

50. Blodgett, O. W., and J. B. Scalzi. *Design of Welded Structural Connections*. The J. F. Lincoln Arc Welding Foundation, Cleveland, Ohio, 1961.

51. Catudal, F. W., and R. W. Schneider. Stresses in pressure vessel with circumferential ring stiffeners. In *Pressure Vessel and Piping Design, Collected Papers*. ASME, New York, 1960.

52. Moody, G. B. How to design saddle supports for horizontal vessels. *Hydrocarbon Processing*, November 1972.

53. Zick, L. P. Stresses in large horizontal cylindrical pressure vessels on two supports. *Welding Journal Research Supplement,* September 1951.

54. Bijlaard, P. P. Local stresses in spherical shells from radial or moment loadings. *Welding Journal Research Supplement*, May 1957.

55. Bijlaard, P. P. Stresses from radial loads and external moments in cylindrical pressure vessels. *Welding Journal Research Supplement*, December 1955.

56. Wichman, K. K., A. G. Hopper, and J. L. Marshen. Local stresses in spherical and cylindrical shells due to external loadings. WRC Bulletin No. 107, December 1968.

57. Leckie, F. A., and R. K. Penny. Stress concentration factors for the stresses at nozzle intersections in pressure vessels. WRC Bulletin No. 90, September 1963.

58. Bijlaard, P. P., and E. T. Cranch. Interpretive commentary on the application of theory to experimental results for stresses and deflections due to local loads on cylindrical shells. WRC Bulletin No. 60, May 1960.

59. Gill, S. S by Ed. *The Stress Analysis of Pressure Vessels and Pressure Vessel Components*. Pergamon Press, New York, 1970.

60. Boardman, H. C. Stresses at junction of cone and cylinder in tanks with cone bottoms or ends. In *Pressure Vessel and Piping Design, Collected Papers*. ASME, New York, 1960.

61. Kraus, H., G. G. Bilodeau, and B. F. Langer. Stresses in thin walled pressure vessels with ellipsoidal heads. *J. Engineering Industry*, February 1961.

62. Kraus, H. Elastic stresses in pressure vessel heads. WRC Bulletin No. 129, April 1968.

63. Gallety, G. D. Bending of 2:1 and 3:1 open crown ellipsoidal shells. WRC Bulletin No. 54, October 1959.

64. Langer, B. F. PVRC interpretive report of pressure vessel research. Section 1. Design considerations. WRC Bulletin No. 95, April 1964.

65. Bijlaard, P. P. Stresses from radial loads in cylindrical pressure vessels. *Welding Journal Research Supplement*, December 1954.

66. Bergen, D. *Elements of Thermal Stress Analysis*. C. P. Press, Jamaica, New York, 1971.

67. Manson, S. S. *Thermal Stress and Low-Cycle Fatigue*. McGraw-Hill Book Co., New York, 1966.

68. Gatewood, B. E. *Thermal Stresses*. McGraw-Hill Book Co., New York, 1957.

69. Bergman, D. J. Temperature gradients in skirt support of hot vessels. ASME Paper No. 62-PET-41.

70. Cheng, D. H., and N. A. Weil. Axial thermal gradient stresses in thin cylinders of finite length. ASME Paper No. 65-WA/MET-17.

71. Wolosewick, F. E. Supports for vertical vessels. *Petroleum Refiner*, October and December 1951.

72. Goodier, J. N. Thermal stress. In *Pressure Vessel and Piping Design, Collected Papers*. ASME, New York, 1960.

73. Cheng, D. H., and A. N. Weil. On the accuracy of approximate solutions for thermal stresses in thin cylinders. *Trans. of the N.Y. Acad. Sci.*, January 1962.

74. Weil, N. A., and J. J. Murphy. Design and analysis of welded pressure vessels skirt supports. ASME Paper No. 58-A153.

75. Weil, N. A., and F. S. Rapasky. Experience with vessels of delayed coking units. *Proc. API*, div. of refining 38-III, 1958.

76. Blodgett, O. W. *Design of Welded Structures*. The J. F. Lincoln Arc Welding Foundation, Cleveland, Ohio, 1966.

77. Bresler, B., T. Y. Lin, and J. B. Scalzi. *Design of Steel Structures*. John Wiley & Sons, New York, 1968.

78. Hall, A. S., A. R. Holowenko, and H. G. Laughlin. *Theory and Problems of Machine Design*. Schaum's Outline Series, McGraw-Hill Book Co., New York, 1961.

79. *AWS Welding Handbook—Section 5. Applications of Welding.* 5th ed. AWS, Miami, Fla., 1973.

80. ANSI Y32.3. Welding symbols.

81. ANSI Y32.17. Nondestructive testing symbols.

82. ANSI B16.25. Butt welding ends.

83. VanVlack, L. H. *Elements of Materials Science and Engineering.* Addison-Wesley Publishing Co., Reading, Mass., 1975.

84. Shrager, A. M. *Elementary Metallurgy and Metallography.* Dover Publications, New York, 1969.

85. Metals Handbook, Vol. 1. Properties and Selection of Metals. American Society for Metals, Metals Park (Novelty), Ohio. Latest Edition.

86. *Low Temperature and Cryogenic Steels—Materials Manual.* U.S. Steel Co., 1968.

87. *Steels for Hydrogen Service at Elevated Temperatures and Pressures in Petroleum Refineries and Petrochemical Plants.* API Publication 941. Latest Edition.

88. *Stainless Steel Handbook.* Allegheny Ludlum Steel Corp., Pittsburg, Penna., 1959.

89. Uhlig, H. H. *Corrosion and Corrosion Control.* 2nd ed. John Wiley & Sons, New York, 1971.

90. Bosich J. F. *Corrosion Prevention for Practicing Engineers.* Barnes and Noble, New York, 1970.

91. Thielsch, H. *Defects and Failures in Pressure Vessels and Piping.* Reinhold Publishing Corp., New York, 1963.

92. Estefan, S. L. Design guide to metallurgy and corrosion in hydrogen processes. *Hydrocarbon Processing*, December 1970.

93. Cooper, C. M. What to specify for corrosion allowance. *Hydrocarbon Processing*, May 1972.

94. Metallurgy for process and mechanical engineers. *Hydrocarbon Processing*, August 1972.

95. Gross, J. H. PVRC interpretive report of pressure vessel research. Section 2. Material considerations. WRC Bulletin No. 101, November 1964.

96. *Aluminum Construction Manual.* Aluminum Association, 420 Lexington Ave, New York, N.Y., 1960.

97. ANSI H35.1. Alloy and temper designation systems for aluminum.

98. Technical booklets available on request from the International Nickel Co., Reynolds Metals Co., Aluminum Company of America, and the Aluminum Association.

99. Holt, M., J. G. Kaufman, and E. J. Wanderer. Aluminum and aluminum alloys for pressure vessels. WRC Bulletin No. 75, February 1966.

100. Weaver, V. P., and J. Imperati. Copper and copper alloys for pressure vessels. WRC Bulletin No. 73, November 1966.

101. Report on the design of pressure vessel heads. In *Pressure Vessel and Piping Design, Collected Papers.* ASME, New York, 1960.

102. Stoneking, J. E., and H. Sheth. Analysis of large saddle supported horizontal pressure vessel. ASME Technical Paper 77-PVP-18.

103. Galletly, G. D. Torispherical shells—a caution to designers. In *Pressure Vessel and Piping Design, Collected Papers.* ASME, New York, 1960.

104. Brummerstedt, E. F. Stress analysis of tall towers. *National Petroleum News*, November 3, 1943.

105. Foweler, D. W. The new analysis method for pressure vessel column support. *Hydrocarbon Processing*, May 1969.

106. Sturm, R. G. A Study of the Collapsing Pressure of Thin-Walled Cylinders. Bulletin 329, Engineering Experiment Station, University of Illinois, 1941.

107. Slember, R. J., and C. E. Washington. Interpretations of experimental data on pressure vessel heads convex to pressure. WRC Bulletin No. 119, January 1967.

108. Williams, C. D., and E. C. Harris. *Structural Design in Metals*. 2nd ed. Ronald Press, New York, 1957.
109. Moody, W. T. *Moments and Reactions for Rectangular Plates*. Bureau of Reclamation, Denver, Colorado, 1963.
110. Taylor Forge modern flange design bulletin No. 502. Taylor Forge and Pipe Works, Chicago, Ill., 1964.
111. Design of flanges for full face gaskets. Bulletin No. 45 Taylor Forge and Pipe Works, Chicago, Ill. Latest Edition.
112. Murray, N. W., and D. G. Stuart. *Behaviour of Large Taper Hub Flanges*. The Institution of Mechanical Engineers, London, 1962.
113. Schneider, R. W. Flat face flanges with metal-to-metal contact beyond the bolt circle. ASME Paper No. 62-WA/PVP-2.
114. Waters, E. O., and R. W. Schneider. Axisymmetric, nonidentical flat face flanges with metal-to-metal contact beyond the bolt circle. ASME Paper no. 68-WA/PWP-5.
115. DIN 2505 (German Standard for Flange Computations).
116. Waters, E. O., D. B. Rossheim, D. B. Wesstrom, and F. S. G. Williams. Development of general formulas for bolted flanges. Taylor Forge and Pipe Works, Chicago, Ill., 1949.
117. Waters, E. O., D. B. Wesstrom, D. B. Rossheim, and F. S. G. Williams. Formulas for stresses in bolted flanged connections. ASME Paper No. FSP-59-4, April 1937.
118. Timoshenko, S. P., and J. M. Gere. *Theory of Elastic Stability*. McGraw-Hill Book Co., New York, 1961.
119. Nelson, G. A. Action of hydrogen on steel at high temperature and high pressures. WRC Bulletin No. 145, October 1969.
120. *Standard Handbook for Mechanical Engineers*. 7th ed. McGraw-Hill Book Co., New York, 1967.
121. Seely, B. F., and J. O. Smith. *Advanced Mechanics of Materials*. John Wiley & Sons, New York, 1960.
122. Boiley, B. A., and J. H. Weiner. *Theory of Thermal Stresses*. John Wiley & Sons, New York, 1960.
123. Houghton, E. L., and N. B. Carruthers. *Wind Forces on Building and Structures*. John Wiley & Sons, New York, 1976.
124. Moody G. B. Mechanical design of tall stack. *Hydrocarbon Processing*, September 1969.
125. *Pressure Vessel and Piping: Design and Analysis, a Decade of Progress*. Volumes I, II, III and IV. Published by ASME, New York, 1972 to 1976. (They contain many technical papers introduced previously as references.)
126. Jawad, M. H. Design of conical shells under external loads. ASME J. Press. Vessel Technol., May 1980.

A9 Abbreviations and Symbols

ABBREVIATIONS

Code	ASME Boiler and Pressure Vessel Code. Section VIII. Rules for Contruction of Pressure Vessels. Division 1.
Code Division 2	ASME Boiler and Pressure Vessel Code. Section VIII. Alternative Rules for Construction of Pressure Vessels. Division 2.
AISC	American Institute of Steel Construction
ANSI	American National Standards Institute
API	American Petroleum Institute
ASCE	American Society of Civil Engineers
ASME	American Society of Mechanical Engineers
ASM	American Society for Metals
ASTM	American Society for Testing and Materials
AWS	American Welding Society
WRC	Welding Research Council
abs.	absolute
C.A.	corrosion allowance
C. G.	center of gravity
cps	cycles per second
Dia.	diameter
°F	degree Fahrenheit
fps	feet per second
ft	foot (feet)
ft-lb	foot-pound
kip	thousand pounds
ksi	kips per square inch
lb	pound (pounds)
mph	miles per hour
mil	0.001 in.
MPY	mils per year
psi	pounds per square inch
psf	pounds per square foot
psig	pounds per square inch gauge
psia	pounds per square inch absolute
rad	radians

SYMBOLS

A, a	area, point, constant
a, b, c, d	distances, points, dimensions
B	point, constant

C	point, constant (usually with a subscript)
D	diameter, flexural rigidity of shells, point
d	diameter
e	unit strain, eccentricity
E	modulus of elasticity, efficiency
F_y	yield strength in column computations
f	load per linear inch
f_w	allowable weld load in lb/lin. in.
h, H	height
I	rectangular moment of inertia
I_w	linear rectangular moment of inertia
J	polar moment of inertia
J_w	linear polar moment of inertia
K	constant, stress concentration factor (usually with a subscript)
k	coefficient, radius of gyration, length
L	length, span in beams
l	unit load in longitudinal direction, lb/in.
M	bending or overturning moment, constant
N	membrane force resultant, number of support legs, grain index
O	point of origin
P	concentrated static load, point, pressure
p	pressure
q	external line load, lb/in.
Q	radial shear, lb/in.
r	radius, radius of gyration
R	radius, reaction force
ΔR	radial displacement
S	direct normal stress, stress intensity, Strouhal number
S_y	yield stress in tension or compression
S_{ys}	yield stress in shear
S_u	ultimate strength in tension or compression
S_{us}	ultimate strength in shear
S_a	allowable stress at design temperature
S_{atm}	allowable stress at atmospheric temperature
S_w	allowable stress in weld
T	torque, thrust force, temperature, period of vibration
t	shell thickness, throat weld thickness
V	wind velocity, transverse force
W	weight
w	weld leg size, unit wind load, unit shell displacement, weight
Z	rectangular section modulus
z	height above ground
Z_w	rectangular linear section modulus of weld
α	angle, coefficient of thermal expansion, numerical coefficient
β	angle, numerical coefficient, concentration factor, cylindrical shell parameter

γ numerical parameter, cylindrical shell parameter
Δ increment, angle in radians
δ deflection
θ angle, angular distortion
λ shell parameter
μ absolute viscosity
ν Poisson's ratio
ρ density, mass per unit volume
σ unit stress in tension or compression
τ unit shear stress
ϕ angle
ψ angle
ω angular velocity
$\propto$ varies as
$\doteq$ approximately equal to
$=$ equal to
$\equiv$ identical
$\gg$ much larger than
∞ infinity
$>$ greater than
$<$ less than
$\leqslant$ less than or equal to

SUBSCRIPTS

a allowable
atm   atmospheric
b bending, bolt
c cone, concrete
e equivalent, earthquake
h head
i inside, initial, intermittent, height level
L longitudinal, larger, linear
n normal
o outside
p due to pressure, permanent, polar
r radial
s steel, shear, smaller, stiffener
t tangential
x with respect to the x-coordinate, height level
y yield
w weld, weight
0 zero

Index